The Last of the Prune Pickers

A Pre-Silicon Valley Story

by
Tim Stanley

2 Timothy Publishing
Irvine, CA

The Last of the Prune Pickers
A Pre-Silicon Valley Story

Front cover photo: Deborah L. Stanley
Other photo credits are at the back of the book

ISBN # 978-0-9842391-2-2
Library of Congress # 2010911003

2 Timothy Publishing
P.O. Box 53783
Irvine, CA 92619-3783
USA
www.2timothypublishing.com

Printed in USA

2 Timothy Publishing books are available at discount for bulk purchases. See web site for details.

For a copy of this book by mail, order on-line or send check or money order to P.O. Box above. Shipping and handling are included in the amount below. Total cost: $24.00 (CA mailing addresses add $1.70 sales tax)

Dedication

In the Madronia Cemetery in Saratoga, California there is a plain grave marker that simply says, "Pitman." No first name, no date, nothing more. It is a fitting memorial for my old friend who did not want to be anybody. But he was. He was a blessing to many, if not all, who knew him. I had the privilege of working for him on his farm for five years, and like many others, I learned a great deal from him. A friend of his put it very well by saying, "He taught me how to think." Yes, and more than that, he set an example for many of us of how to live. This book is dedicated to the memory of Robert Freeman Pitman.

The Santa Clara Valley and Vicinity,

California

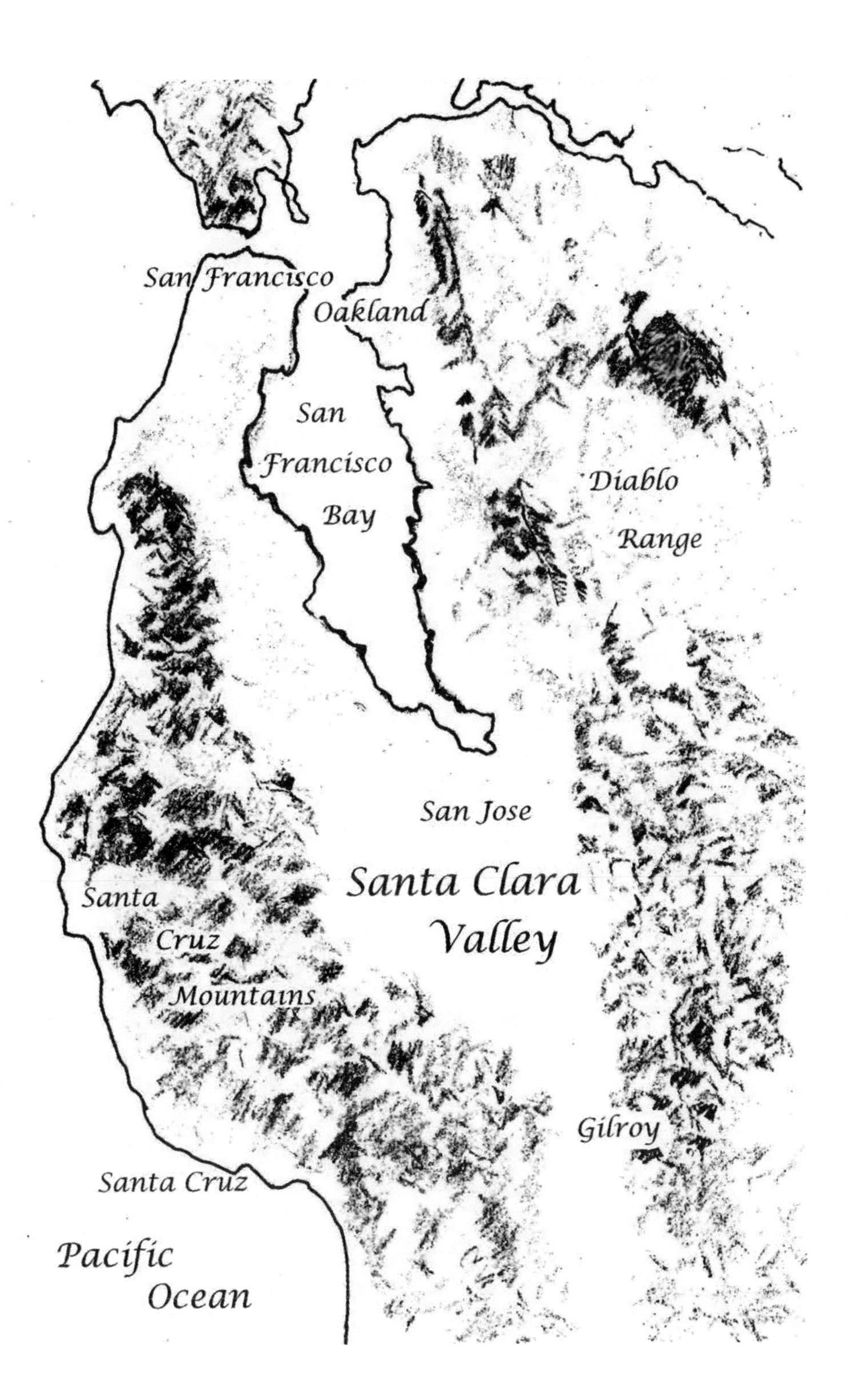
San Francisco
Oakland
San
Francisco
Bay
Diablo
Range
San Jose
Santa Clara
Valley
Santa
Cruz
Mountains
Gilroy
Santa Cruz
Pacific
Ocean

Of Plums and Prunes

Prunes are plums, I won't deny,
But not all plums are prunes.
The kinds that ferment—those are plums;
Those that resist are prunes.

Plums are picked from ladders,
From branches of the tree;
Prunes are picked up off the ground.
This differs too, you see.

Plums are to be eaten fresh;
Prunes are grown to dry.
I hope this explanation helps—
At least I had to try.

Now if you don't believe me,
With my grammar can't agree,
Mr. Webster understood these things,
And he will vouch for me.

Acknowledgements

First, I must give thanks to God who allowed me to meet Robert Pitman and work on his farm when I was a young boy. It was a good way to grow up, and I will always be grateful for that privilege.

Thanks to my wife, Deborah, for her encouragement, support, attentive ear, and help with the photos. After my first book, *Letters to My Feathered Friends*, I didn't know how she would feel about me embarking upon another. My concerns were unfounded. When I told her about this book that had formulated in my mind, she was overjoyed. She shocked me then and I'm amazed that, after all these years together, she still wants to hear what I have to say.

Thanks to Dave Pitman, Robert Pitman's younger son, not only for being generous with his time in sharing with me about his dad, but also for telling me many stories about growing up on the farm, and for supplying some of the photos for this book. Thanks to Bob Pitman, his older brother, for sharing with me his experiences, and to Tom Pitman, Dave's son, for sharing his.

Thanks to Ernie Kraule, a former Pitman prune picker, who told me of his time on the farm and who supplied one of the photographs. Thanks to George MacKenzie, another Pitman farm alum, who told me what he remembered. Thanks to Frank Matas and his wife, Lee, who were close friends of Robert Pitman, and shared something of their experiences with me.

Thanks to Sara McCullough, my aunt, for sharing with me some of her experiences of working in a cannery.

Thanks to Annette Stransky at the Saratoga Historical Foundation, Russell Skowronek at the University of Santa Clara, Charlene Duval at Sourisseau Academy for State and Local History, San Jose State University, Susan Snyder at the Bancroft Library, University of California, Berkeley, Stacy Mueller and Lauren Miranda Gilbert at the San Jose Public Library, Mary Hanel at the Santa Clara City Library, Sara Kempen at the Sunnyvale City Library, Chris Wright and Patrick O'Hallaran at the Irvine Heritage Park Library, Michael

Griffith at Santa Clara County Archives, Sara Puckitt at History San José, the Sunnyvale Historical Society staff, the Saratoga Library staff, Lisa Christensen at the California History Center and Foundation at De Anza College, David Rumsey of Cartography Associates, Michael Fossati at the City of Saratoga, Rich Peterson and Nancy Smith at the California Dried Plum Board, Stella Mentink, Sharon Brown, Bruce Norton, and Mark Dalrymple at Sunsweet, Bob McClain at the California Pear Advisory Board, William Ferriera at the Apricot Producers of California, Franz Niederholzer, Kent Bradford, and Joseph Ditosaso at the University of California, Davis, Carolyn Vogt Artru Feroben at mariposaresearch.net, and Mike Kealy at Kealy's Korner Books, all of whom were helpful and supportive.

Thanks to the farmers who helped: Ken Lindauer, Marlene Orsetti Manzo and her husband Don, Charlie Olson, and George Novakovich.

Thanks to Casa de Fruta, in Hollister, California, for preserving and displaying their excellent collection of antique farm machinery.

Thanks to all who contributed the use of their photos for this book. The book would be lacking without each one.

Thanks to our daughter, Jennifer Stanley, to Bruce Giddens, Becky Hatley Reed, and Rebecca Twitchell for their editing work. You all made me look like a better writer than I am. Any remaining errors or awkwardness in the text are mine.

Thanks to our son, James Stanley, to Reed Hazen, and to Tricia Stanley, for their helpful comments and suggestions.

This book could not have been completed without the help of all of you. Thank you!

Orchard and foothills, Santa Clara Valley, Cal.

Preface

Not long before silicon reigned in the Santa Clara Valley of California, the Valley was largely covered with orchards. There were orchards of pears, apricots, cherries, walnuts, and the king of them all: prunes. Thousands of the orchard farms were small family operations. This is the story of what preceded those farms, how they came into being, and how they thrived. It is also the story of one of the last of those farms, of the farmer, and of some of the young boys and girls who had the privilege of working for him.

The book is written in two parts. Part One is a well researched, yet engaging, general history of the area. The emphasis is on agriculture, particularly on the rise and decline of the orchard farms. Incorporated into it are many important pieces of the historical picture that are lacking in many, if not most, historical accounts. I have tried to present the "*how*" and "*why*" behind the events that took place. The history is well documented with references and notes at the back of the book. Many of the notes are interesting stories in themselves.

Part Two narrows the story down to one farm, which was typical of many of the Santa Clara Valley orchard farms. It gives some of the history of that farm and of the people who lived and worked on it.

In both parts, I endeavored to present many lesser-known matters of importance: the applied physics, chemistry, and life science behind the early industries, some of the key principles involved in farm practice, and a look into the mindset of a farmer.

Throughout the book, I have also attempted to point out some things of lasting value. I hope you find this refreshing.

The Last of the Prune Pickers
A Pre-Silicon Valley Story

Contents

Part One

A Short Agricultural History Presented in an Uncommon Way

1
Beginnings

At the southern end of the San Francisco Bay lies a valley that stretches roughly fifteen miles from east to west and thirty miles from north to south. That valley, called Santa Clara, was so fertile that for some time it was described as "The Garden of the World." Although promoters tend to exaggerate, if you ever had the privilege of looking down on the Valley from Blossom Hill Road in Los Gatos on a spring day sixty years ago, you probably would have been inclined to agree with them. To complement the fertility of the land, the climate of the Valley is most comfortable. The temperature rarely goes below 30 degrees or above the mid 90s, and an afternoon breeze coming off the bay serves as natural air-conditioning making for pleasant, cool summer evenings. To top it off, the Santa Cruz Mountains, with their fragrant mixed forests of oak, madrone, bay, and redwood trees lie just to the west and south.

The history of the Santa Clara Valley is closely tied to the history of California as a whole. Unlike in Mexico and in the Andes of South America, there were no great civilizations in California. In fact, the peoples of California were the antithesis of such civilizations, and the people inhabiting the Valley were typical of them. Before the coming of the Spaniards, the Valley was inhabited by a people who are now referred to as the Ohlone. The Spanish called them the Costaños, *The Coastal People*. These people, and the other tribes in California, were not organized into large tribes or nations as were most of the native peoples in what is now the United States.

The Ohlone (usually spelled Olhone in older books) were divided into about fifty tribal clans who lived in the San Francisco Bay area and extended southward into the Salinas Valley. They spoke no less than eight dialects of a common language, intermarried, and traded with neighboring clans, but did not consider themselves part of a group larger than their own clan. They were known by different names such as Tamien, Puichon, Lamchin, and Muwekma, which simply means, "the people."[1] Each group, usually consisting of between fifty and a few hundred members, had its own name, leader, and customs. Some of their villages were of a seasonal nature, as many Valley dwellers would go up to the mountains and even to the ocean during different seasons. They went to the mountains and to the ocean not only for food, but like those who came after them, simply to be refreshed. The Ohlone lived a simple life of hunting and gathering their necessities. They fished, using nets; hunted game, using bow and arrow; and gathered roots, seeds, berries and most predominantly, acorns. Acorns were the primary food of many, if not most, of the tribal groups. They beat the acorns into a course meal and made cakes of it. By grinding the meal in stone mortars with stone pestles, bits of sand were released into the meal that caused their teeth to wear down quickly. As a result, they tended to live very short lives.[2]

There is no record that these early Santa Clara Valley dwellers displaced others before them, but that they defended their territories and were from time to time engaged in armed conflict with their neighbors is well documented.[3] If the rest of human history is any indication, we can safely assume that they displaced one another, whether by persuasion or by force, whenever the desire for what someone else had and the strength to take it away from them arose.

Huge treasures of gold and silver were found in the great civilizations of the Aztecs and the Incas, and the Spaniards, being possessed of greater military ability, helped themselves to all of it. Juan Cabrillo had discovered California and claimed it for Spain in 1542, but there appeared to be nothing worth taking, so the native peoples were left alone for over two hundred years until the mining operations in Mexico and in the Andes started to play out.

The British, who in 1579 at the hand of Francis Drake had also made a feeble claim on California, by the mid 1700s were desirous of expanding their empire and renewed their interest in California. By that time also, the Russians were proceeding further and further south, following the fur trade, and had established forts on the West Coast as far south as Fort Ross, which is about eighty miles north of San Francisco. This forced the original claimants of California, the Spanish, to act.

The treasures that the Spanish had stolen from the Incas and the Aztecs had been depleted by two centuries of wars waged trying to protect them. This being the case, the Spanish faced the problem of how to control California with limited resources. The solution they decided upon was to transform the native peoples into loyal Spanish citizens. Toward this end, the Spanish began their occupation of Baja, or "Lower," California and by the mid 1700s had established a few missions there. The Spanish occupation of Alta, or "Upper," California began with the Portola Expedition of 1769.

The Portola expedition consisted of four detachments. Two overland parties departed from Mission San Fernando de Velicata in central Baja, California and two ships sailed from La Paz, a port on the eastern side of southern Baja. Father Junipero Serra, the priest charged with establishing the mission system in California, came in one of the overland parties with Gaspar de Portola. Using maps of the coast that were a hundred and sixty-seven years old, the four parties were all to converge on San Diego at about the same time. The detachments all arrived, but not without losses. About half of the 219 men in the expedition died from the journey or shortly after it—most from scurvy. From that time, and for the entire duration of Spanish and Mexican rule, it was difficult for these governments to get anyone interested in settling in California. But the expedition was successful from the standpoint that a presidio, or fort, and a mission were established in San Diego.

Of even greater importance, San Francisco Bay was discovered during the later part of the Portola expedition, and with that discovery California became the coveted prize of the European powers. The importance of the bay—one that could contain all the navies of Europe—could not be overstated. It was the only safe harbor north of San Diego on the California coast.

The discovery of San Francisco Bay gave Spain a heightened sense of urgency in occupying California. Over the next thirty-five years, the Spanish established nineteen missions in Alta California,[4] all close to, or on the coast. As fertile land and a substantial Indian[5] population allowed, the missions were spaced about a day's journey apart. At the same time, pueblos, or towns, were established. The first two sites selected for the pueblos were well chosen—they eventually became San Jose and Los Angeles.

To protect the missions and the pueblos, the Spanish built four presidios in California. Each contained a detachment of soldiers. The presidios were located at San Diego, Santa Barbara, Monterey and San Francisco. They served not only to protect the missions and the pueblos from occasional Indian attacks, but just as importantly to keep out foreigners. Thus the Spanish began to occupy the land they had claimed 230 years earlier by sailing up and down its coast.

Due to the difficulties of sea travel, the Spanish needed a direct overland route from New Spain (Mexico) into California. In 1774, Juan Batista de Anza was given the assignment to find that overland route. He did, and two years later, in 1776, he led a small party of pioneers overland into northern Alta California. The expedition departed from Sinaloa and Horcasitas, in what is now the state of Sonora, Mexico, and consisted of 242 soldiers, priests and settlers. These people formed the nucleus of the mission and presidio in San Francisco, Mission Santa Clara, and Pueblo San Jose.

The overland route into California established by Anza was not an easy one. Besides going through the Sonoran desert, the route crossed the Colorado River, and that meant going through Yuma territory. The Yumas had important villages at the fords of the river and they, unlike the other California Indians, understood warfare.

Anza acted with intelligence. He gave the Yumas presents, treated them with respect, and in return was given safe passage. Coming after him a few years later with the intent of establishing towns and a mission at the fords of the river, other Spaniards were not as smart. They helped themselves to the best of the land, let their livestock graze on the Yuma's tilled land, occasionally flogged a disagreeable Yuma, and even put their leader in stocks because he did not yield to their every wish. The Yumas were not appreciative, waited for an opportune time, and responded in kind. They burned down the Spanish settlements, killed several of the inhabitants, and for all practical purposes closed the door to overland travel into California for the remaining sixty-five years of Spanish and Mexican rule. [6]

Photo: Native Californian women ground acorns into meal on large flat rocks such as this one. In time, the hollowed out mortars were formed.

2
The Rise and Fall of the Missions

Mission Santa Clara was founded in the heart of the Santa Clara Valley in 1777 by members of the Anza Party, and due to its fertile location it became one of the foremost of the California missions. The site that the Anza Party members selected for the mission was in the midst of several Ohlone villages along the Guadalupe River. Two miles up the river, on the opposite bank, other members of the same party founded Pueblo San José at about the same time. Exactly how the Ohlone felt about the Spaniards moving in is anybody's guess. But they did not repel them.

For the first several years, both the mission and the pueblo had a very rough time. They could not support themselves and were dependant upon supply ships from New Spain (Mexico) even for their food. And to supply them was not easy. During the Spanish and Mexican eras, the departure port for the supply ships was San Blas, which is just north of Puerto Vallarta. The voyage is only about two thousand miles, but the winds and current are contrary all the way. By the time these sailing ships had tacked back and forth up the coast, a few months had passed and outbreaks of scurvy were common. As happened to the Portola Expedition, many who undertook the journey died.

The maintenance of the mission and the pueblo was a significant drain on the Spanish crown. To complicate matters further, the settlers soon discovered that the Guadeloupe River flooded from time to time, and both the mission and the pueblo had to be moved to higher ground. The mission was moved within the first few years; the pueblo needed to get flooded out a few more times, and was finally moved in the early 1790s. To live in the floodplain with easy access to the river was convenient but precarious, so to survive, the newcomers tended to live and plant crops wherever they could.

The mission log books, which catalog human population, marriages, deaths, etc., use two terms to describe human beings: first there were the "Men of Reason," then the "Indians." It is interesting to note that the "Men of Reason" did not learn from the Indians how to live in the new land. They did not learn to find edible foods as the native peoples did. The Indians throughout all of California had learned how to transform the naturally toxic acorns from the local oak trees into their daily bread. They cracked the shells to extract the meat, ground it into a course meal, and then carefully rinsed the meal to remove the tannic acid. They placed the ground meal in expertly woven watertight baskets, added water, and brought it to a boil by adding hot rocks. They then poured off the water until the cloudy water ran clear. Europeans who tried to eat the acorns without going through this process became quite sick. Likewise the "Men of Reason" did not have the good sense that the Indians had to move to higher ground during the rainy season.

In the long run, however, the newcomers prevailed. Their crops did well when they were not flooded out and Indian labor was available to tend to them, and their livestock, needing very little care, multiplied very rapidly. By 1792, Mission Santa Clara's few cattle had increased to over 4,000, the sheep to 800, and the horses to over 600. These, along with the livestock from the pueblo, gobbled up the acorns and pushed out the deer and elk that the Indians depended upon.

In the beginning of their stay, the mission fathers, or padres, established good relations with some of the natives by giving them presents and food. The children were more trustful than the adults, but as the Indians' food became more and more scarce, the hearts and minds of the adults were persuaded by their stomachs. Within a decade, the mission population had reached 647. With the drought years of 1794 and 1795, tribal society completely collapsed and the mission was swamped with hungry Indians. What the padres could not do by persuasive speech and gifts, their livestock had done for them. The Indians, defeated and depressed[7], were "converted" and embraced their new fathers.

A new communal life that included all the local Indians thus emerged. No longer were there Indians in the Valley that were not a part of the mission.[8] All were put to work building a large mission village that was to supply all their needs. They ate together, lived in dormitories and attended mass daily.

It could be said that the mission was primarily a school—a new way of survival needed to be learned—and it was run like a military camp. Food, and the ability to produce it, was there and that made for a peaceful atmosphere most of the time. Of course, there were occasions when some of the padres or Indians behaved in a manner detrimental to the good of the commune. Sometimes a priest pushed too hard—one in particular nearly caused a revolt before he was advised to, "Use less zeal." Sometimes an Indian, or group of them, deserted or otherwise "misbehaved." But overall there was peace, as is proved by the fact that for nearly the entire life of Mission Santa Clara only a handful of guards was needed to keep order over a population of more than a thousand "neophytes."[9] This kind of peaceful existence was not the case at the nearby Mission Santa Cruz or at several of the other missions.

If you think you have a difficult job, consider that of the mission priests. They were to serve as diplomats, ethnologists, farming experts, scientists and geographers. They were micro-managed from afar by people who had no understanding of local conditions. They were to keep fastidious records, produce income for the crown, get along well with the neighboring pueblo citizens (many of whom were notorious for drunkenness and chasing Indian women), were to please both the church and the state, and be mild in the discipline of the serfs.[10] Oh, and they were to change the entire way of life of a people who had been in the land for a thousand years, and make them into loyal Spanish citizens and good Catholics—even though many from their own culture set a very poor example.

The reason for the success of Mission Santa Clara can be stated in one word: food. The Indians' former food supply was no longer available in sufficient quantities, and the newcomers knew how to pull food out of the fertile valley. Thus Mission Santa Clara became one of the most successful of the California missions. At its peak, it was populated by about 1500 Indians, a handful of priests, and a few families of Spanish[11] descent that served as foremen, artisans and "escorts" or guards. At the beginning, most of the Indians lived in the villages around the mission. Later, huts and dormitories were built on the mission grounds. With Indian labor, in a very short time a thriving agricultural commune emerged.[12]

However, there was a big problem associated with the building up of the missions—the Europeans brought their diseases with them. The Indians did not have immunity to these foreign diseases, and over time the great majority died from them. So at the same time the enterprise was being built up[13], the labor force was being diminished. To keep the mission functioning, the Spanish went into the interior and rounded up more Indians.[14] This was the beginning of the end of the mission system in California, and when coupled with increased monetary demands upon the missions by the Spanish government, within just a few decades the system began breaking down.

Pueblo San José (not to be confused with Mission San José, established in 1797 about 12 miles to the northeast) was the first Spanish civil settlement in Alta California. Nine soldiers with their families and five additional settler families from the Anza expedition originally settled the pueblo. They were conditionally granted small home sites in the pueblo and about ten acres of farm land just outside the pueblo. In addition to land, the pueblo settlers were given a loan of clothing, farming tools, and two each of oxen, mules, cows, mares, sheep, goats, and pigs. To secure a deed to the land granted, they had to remain for a minimum of five years. Pueblo San José, which later became the city of San Jose, was located just south of today's San Jose International Airport. It was established as a farming community, and was to provide food for the presidios. In actual practice, both the missions and the pueblo served this function. The missions, with abundant cheap labor, were by far the more efficient farming operations. The population of the pueblo, on the other hand, did not grow much during the Spanish era. Retired soldiers settled there, but few families came.

The mission and pueblo settlements could hardly have been more different—and neither appreciated the other's presence in the Valley. In the beginning, Father Serra pleaded with the central government to overrule Governor Neve's decision to establish the pueblo in San José, as he felt it would be detrimental to the mission. His pleas were not heeded, and during the entire mission period there was constant bickering between the civil government and the missions over boundaries, cattle grazing, cheap labor, and that most valuable of all California assets—water.

At the height of the mission system's success came the Mexican war for independence from Spain, which lasted about twenty years. After the war began, local officials were required to sign documents of allegiance to the tottering New Spain, but otherwise California was not involved. There was one notable exception. Toward the end of the war, in 1818, there were destructive rampages in Monterey, Santa Barbara, and San Juan Capistrano by the French pirate, Hypolite Bouchard. Bouchard preyed upon Spain's weakness and helped himself while supposedly helping the Argentines in their own war for independence.

When the war for independence was finally over, new officials came with new papers of allegiance to sign. This was done without too much trouble—the civil officials evidently understood that it probably would not make much difference to them anyway, and the mission padres eventually accepted the obvious.

Although from the very beginning of Spanish rule there was criticism of the mission system, the criticism increased greatly under Mexican rule. Mexican independence from Spain came in 1822, and with it an all-time low in the stability of the mission system. Indian desertions became frequent, hostilities increased, and in general there was an attitude of revolt. After many changes of government and much vacillation on the matter, the missions were secularized beginning in 1833. That is to say, the Indians were "freed" and the mission lands were taken over by the Mexican government. Mission Santa Clara was taken over in 1836. It is worth noting that at the time of the takeover only a third of the remaining mission Indians were Ohlone. The rest were Yokut and Miwok from the Central Valley.[15]

3
The Californios

In the last chapter, it was stated briefly that from the beginning of the Spanish occupation of Alta California there was an unpleasant rivalry between the missions and the secular interests. Nowhere was this more intense than in the Santa Clara Valley between the pueblo and the missions. Both complained bitterly against the other to the central government. Mexican Independence intensified these disagreements considerably, and with the coming of the Hijar-Padres colonists in 1834, this friction turned into a maelstrom.[16]

Hijar

José Maria Hijar and José Maria Padres were wealthy men of great influence in the Mexican Government.[17] Spurred by a law passed in 1833 requiring the immediate secularization of the missions, and Hijar having been given a special commission to regulate their secularization, these men gathered to them about 250 people to immigrate to California. Each family was offered a $10 signing bonus, transportation to San Blas, meals and passage to California, a farm from the newly acquired public land, and rations for a year to get them started. Even livestock and tools were thrown in to sweeten the deal.[18] In other words, they were to receive at least as good a deal as the earlier colonists had. From the time of Portola, such were the desperate measures the Spanish and Mexican governments went to in order to get anyone interested in settling in California.

Before the Hijar-Padres party arrived however, a new government had been installed in Mexico, and the new president, Santa Ana (of Alamo fame), canceled Hijar's commission while the party was en route. Meanwhile, one of the two ships carrying the colonists was nearly lost in a gale off of Point Conception. The other ship was considerably faster and was anchored in Monterey Bay when the same storm hit. The storm drove the ship onto the sandy beach and the surf broke it apart, claiming the lives of three people on board. After these disasters, the surviving colonists arrived in Monterey to find out that all of the government promises of land, etc. were canceled. What started out as a promising venture had quickly collapsed. The politics took a while to sort out, and instead of many from the party establishing a colony in the Santa Rosa Valley, as was originally planned, the colonists spread out, mostly over the northern part of Alta California.

The people Hijar and Padres brought with them were of a much higher class than all but a few who had come to California previously. Many were educated, some were property owners, all had a trade or profession, and according to historian Hubert Bancroft, there was also the "noticeable absence of criminals among them." This was in stark contrast to many who had come before them. Until that time, California had been largely a retiring ground for military personnel, and sending prisoners there—political or otherwise—was not rare.[19]

The local aristocracy's reaction to the new colonists, however, was not favorable. The Californians, or "Californios" as the people were called, had received a group of colonists made up of teachers, artisans, craftsmen, and farmers—exactly what they had been asking for—but it appears that the entire larger San Francisco Bay Area was not big enough for both the newcomers and the existing aristocracy.

The whole matter was in fact an ugly political fight. One side wanted to do away with "missionary infestation," the other wanted to maintain the status quo. Mariano Vallejo, whose huge land holdings were closest to Santa Rosa, and others of the aristocracy, claimed that the newcomers were going to plunder the missions and even secede from Mexico. In presenting their cases to the central government, both sides used rhetoric concerning the betterment of the masses of the people, and both expressed great idealism when it came to the Indians. The real issue, however, was land and social status. And who was going to get the plunder from the missions.

Land Grants

After the break-up of the missions, the Mexican government tried to encourage settlement in California by offering land grants. Land had previously been granted to settlers, by both the Spanish and Mexican governments, but not nearly to the degree as after the confiscation of the mission lands. It was land that lured the Hijar-Padres settlers and those after them. The land grants fell into three categories: building lots, called *solares,* in the pueblos; small farm plots of about ten acres, called *suertes,* near the pueblos; and large ranches, or "ranchos," which were usually at least several thousand acres. There were also common lands, called *ejidos*[20], for pastureland and wood.

During Mexican rule, there were forty-one grants given for rancho property in what is now Santa Clara County. These grants were given between 1833 and 1846, and were in addition to the three large Spanish land grants that were already in existence. Rancho grants eventually included most of the land in the Santa Clara Valley, and some of these grants went to Hijar-Padres colonists.

At the time these grants were given, all of California was considered cattle country, and as such, most of the rancho grants were of several thousand acres or more. The notable exceptions in the Valley were the Bennett and Enright tracts and the Embarcadero[21] de Santa Clara. The Bennett and Enright tracts were near the mission and were 358 and 710 acres respectively. Both included cultivated mission land. The Embarcadero de Santa Clara, which was the bay waterfront between Santa Clara and present day Alviso, was at the mouth of the Guadalupe River. All through the Spanish and Mexican eras, cowhides were loaded onto barges there to take them to ships anchored in San Francisco Bay. These three smaller parcels were granted late in the period, in 1845.

With the break-up of Mission Santa Clara, about half of the mission land—which encompassed a large and disputed portion of the Valley—was to be given to the Indians. Some of it actually was. Three rancho grants were given to mission Indians, two of them from mission lands.[22] One of these, near or on the present day Moffet Field Naval Air Base, was given to a man named Inigo (or Ynigo), who kept his Indian name. The other Indian grantees had Spanish names. The Indians who had received land grants, like the other grantees, were to work the land and pay taxes or abandon it. Most of the Indians from Mission Santa Clara chose not to work the land and fled into the interior, which until the coming of the Americans was still Yokut and Miwok territory. By 1839, only about 300 Mission Santa Clara Indians remained in the vicinity of the mission. Those who remained in the Valley worked on the

ranchos in much the same way as they had on the mission.[23] Sixty plus years of the mission system had produced very few Native Americans who were truly assimilated into Hispanic culture.
In order to receive a grant of land in California, a person needed to be a Mexican citizen or become one, have a good reputation, and pledge allegiance to Mexico. The rules concerning who was eligible to receive a grant varied somewhat over time and through changes in government, both in Mexico and in Alta California. Generally speaking, to receive a grant, one needed to petition the governor with a description of the parcel desired, file a hand drawn map, or diseño, of the area, and submit a fee. The petition could be for a pueblo lot, a small plot of farm land, or a 30,000 acre rancho. For a rancho, one needed to show the ability to put livestock on the property. The governor, if he wasn't replaced before he reviewed the petition, would make a decision, and if he approved, would issue a formal grant in writing. There were surprisingly few takers from Mexico, and before the flood of Americans began, a few Europeans and Americans were given grants. The perils of the sea voyage and apprehension about living in a society greatly outnumbered by Indians apparently deterred most Mexicans from coming.

The number of people who filed for rancho grants in the Santa Clara Valley and were rejected is unknown. What we do know is that some of the grants went to heirs of Anza expedition members, many went to soldiers who were usually of high rank, many went to government officials, at least two went to those who had helped oust previous administrations, two went to Californio women,[24] one went to a British consulate, one went to an American[25], and as mentioned previously, three went to mission Indians. In all probability, some simply went to the highest bidder. The rancho grants after the break-up of the missions issued in the short-lived era of the Californio "Dons." The Dons, or land barons, were the rancho owners, or rancheros. They were from a relatively few wealthy families, and many owned multiple ranchos. They represented about five to ten percent of the non-Indian Californio population, controlled the government, and tended to intermarry.[26] Appendix II is a detailed chart of the Santa Clara Valley rancho grants.

Trade

It is estimated that in 1830, there were less than one hundred non-Spanish/Mexican foreigners in all of Alta California, and probably no more than four in the Santa Clara Valley area. Spain's policy, from the beginning of their involvement in the Americas, had been to keep all foreigners out of their territories. For the entire 300 years of the empire, foreign trade was officially, though not practically, forbidden. The Mexican government opened trade, but required all who wished to trade to obtain approval and pay taxes in Monterey first. Of course, not all did. And not all cared.

California was cattle country, and trade on the coast was for cow hides and tallow.[27] Ships came from Europe, the Americas, Russia, the Sandwich Islands (Hawaii), and elsewhere to buy California hides. The US trade was completely dominated by a few merchants in Boston. Nowhere in California were there more hides shipped out than from the Santa Clara Valley and the surrounding area.[28]

As mentioned previously, the cattle multiplied very rapidly in California. It was perfect cattle country and the longhorn cattle (the same breed as "Texas Longhorns") required almost no attention at all except branding and castrating. Bulls are competitive for the ladies, steers don't care; so the males were castrated, not to make their meat tender as is done today, but to keep them from fighting over the females. In other words, to keep them from killing each other before it was time to take their hides off of them.

During the Mexican era, many Europeans and Americans came to California. Some had agreements with their employer to remain there, others were discontented sailors who had deserted their ships. Although desertion from a merchant vessel was a major offense, many sailors took the risk and found a haven with the Californios who were starved for European-based companionship. Most of those who came married Californio women and became part of Californio life. Such was the case of the first American settler in the Santa Clara Valley, Philip Doke, who, in 1822, left a whaling ship and settled near Gilroy on the rancho of Mariano Castro and married one of Castro's daughters. Another, a Scotsman named John Cameron, who changed his name to Gilroy, left a ship in Monterey and eventually married into the wealthy Ortega family.[29] John Burton, from Provincetown, Massachusetts, came around 1825 as master of a ship that was wrecked off of San Diego.[30] He settled in

San Jose. Burton was well respected by the Californios and later became a US judge in San Jose. These, and all other American settlers in the Valley that we know of in the 1830s, came by ship. With the freeing of the Indians, the sailors had to do more of the work with the hides than they had done previously, and this enabled many more of them to come ashore and see more of the countryside. Understandably, they became quite fond of it, of the way of life, and of the ease and plenty that was everywhere. An increasing number of sailors deserted their ships and stayed.

Way of Life

As the missions faded away, Pueblo San José grew. The pueblo population was about 285 when the Mexican government took over in 1821,[31] and grew slowly until the 1840s. Depending on which historian you read, by the time of the American conquest, the pueblo was either a ramshackle affair or a thriving town of 700 residents. It was probably both. The residents were a mixture of Mexicans, Chileans, Peruvians, Europeans, and in the later years, Americans. The Chileans and Peruvians had been recruited by the Mexican government late in the period in a desperate, but unsuccessful, attempt to balance the influx of American settlers. Until perhaps the middle 1830s, most of the pueblo homes were huts made of willow branches and covered with thatching, similar to those that the Ohlone had built. The main difference in construction was that those of Spanish descent plastered their dwellings with mud.

The Californio way of life before the American occupation was, by all accounts, laid back. For the men. Some of the Californio men worked as subsistence farmers on their own plots (suertes) or on the common lands (ejidos), but the great majority of them worked for the rancheros. Commonly the men worked just twice a year during the cattle round-ups, which were to brand and castrate, and to butcher. The Californios' one industry was to raise cattle for hides and tallow, and these they traded for what they wanted. What they wanted included shoes and other leather goods made on the East Coast from the hides that they had shipped there, and they were quite content with this arrangement.

As for the Californio women, much has been written about the beauty, social graces, and hospitality shown to visitors by the women of the ruling class. There is no need to repeat those things in detail here. But for the 90-95 percent of the women who were not in that class, life was quite different and very little has been recorded. The story of one of these women, however, gives a small window into the life of many of the women who made up the vast majority of Californio women. That story is included in a note on Californio women referred to earlier in this chapter.

A little explanation may be helpful here as to the difference between rawhide and leather, and to underscore the disposition of the Californios at the time. Making leather from rawhide is done by a process called tanning. A cleaned and dried hide is soaked in a vat containing tannin, an acidic chemical compound. Tannin alters the chemical protein structure of the animal hide and the resulting leather does not decompose easily.[32] The hills to the west of the Santa Clara Valley are full of Tan Oak trees, which because of their tannin-rich bark derive their name from this process. Although the equipment needed to make most leather is modest, apparently the Californios were not interested in developing their own leather industry, or any other industry for that matter, so they shipped raw hides.

The Californios' superb horsemanship skills, as well as their distain for other kinds of work, were common themes in the writing of visitors or newcomers to California. One of the early American pioneers, John Bidwell, of Chico fame, wrote in his journal, "It is a proverb here, and I find it a pretty true one, that a Spaniard will not do anything which he cannot do on horseback."[33]

The ships that came to California for the hide trade were floating department stores. When a ship came in, it was a grand affair and the captain invited the local people on board. They came, did their shopping, and paid for what they wanted with cow hides, commonly known as "California Dollars," and tallow.

The period of the Californio Dons has been greatly romanticized by novel writers and Hollywood filmmakers, but there is little romantic about butchering cows and leaving their carcasses to the dogs and vultures. In fact this—not expansive vineyards and happy mariachis singing to beautiful señoritas—was the way of life at nearly all of the ranchos.

However, the climate and fertile soil allowed the Californios to live without expending much effort, and this time of ease could not last long. Word got out to places east where people lived a much harder life. Fur trappers who came from the north, and hide traders who came by sea, brought the news east of California's fertile valleys, mild climate, easy life, and all but deserted coast. Added to these reports was the great popularity of the book, *Two Years Before the Mast* by Richard Henry Dana. Dana's book had the same effect as the televised Rose Parades had a little more than a hundred years later, and many folks in the eastern states began to develop an itch. The itch became talk, and by the 1840s the talk had become action for many.

The Cinderella time for a few wealthy California land owners was over.

4
The Americans

Although there were few Mexicans interested in settling in California,[34] there were plenty of interested Americans. By the time it became known that the overland route into Oregon Territory was feasible by wagon, Mexico's rule was all but over.

From the opening of the Oregon Trail in 1841, Americans[35] began to immigrate into Oregon, each year in larger numbers. Although the great majority of these people settled in Oregon Territory, some came down into Alta California. At that time, Oregon Territory extended from British Columbia down to about 80 miles north of San Francisco and included much of what today is Northern California. The territory was under joint British and American jurisdiction, though it would not remain so for long.

Some settlers who were headed to Oregon changed their mind on the way because of something they heard and headed south into California.[36] Of those, many came into the Santa Clara Valley. One of the early parties into the Valley was lead by a trapper named Elisha Stevens of Stevens Creek fame. The party came in 1843, and included the Murphy family. The patriarch, Martin Murphy, purchased Rancho Ojo de la Coche, near today's Morgan Hill, from Juan Maria Hernandez in 1845. In a little more than a decade, the Murphy family controlled six of the largest ranchos in what is now Santa Clara County.[37] Several other Americans also purchased land from Californios in the mid 1840s. According to Mexican law, one needed to be, or become, a Mexican citizen before they could purchase land in California, but it appears that at times this requirement may not have been binding.

The new immigrants to California were very different than those who had come before. The previous American immigrants had come from the eastern seaboard, all were involved in trade, and until the late 1830s, if they decided to stay, they blended in with the Californio culture. They married Californio women, became Mexican citizens, and as such embraced the Catholic Church. The new immigrants were nearly all from the frontier states, that is, west of the Appalachian divide, and had no desire to become either Mexican citizens or Catholics.[38] They, or their fathers, had gone pioneering before, and to them, coming to California was no different than their previous migrations.

Virtually all of the early immigrants, who took the Oregon Trail on their way to California, visited Sutter's Fort at what later became Sacramento. Sutter, from Switzerland, had received a grant for a large tract of land there in 1841, and had established a town and trading post. There the newcomers were warmly received and got their orientation to California from Sutter's perspective.

Most of the new immigrants settled in the Central Valley—some in Oregon Territory, some in Mexican California. It should be noted here that neither the Spanish nor the Mexicans ever had control of the interior of California. Though they claimed it, other than Sutter's establishment and a few ranchos, it was virtually unsettled.[39]

Even before the Oregon Trail was opened, the Mexican authorities, both in California and in Mexico, realized that the Californios could not hold up to the pressure of the American immigration. Deserters from merchant ships had become much more numerous during the 1830s, and by the end of that decade many were arrogant and not as kind to their hosts as those who had come before them. According to Bancroft, many of these were lawless men, typified by Isaac Graham,[40] himself a trapper, who eventually settled on the coastal side of the Santa Cruz Mountains.

Graham, and those who gathered to him, despised the Mexicans and were bent on rebellion against them. They were also well armed and some were expert riflemen. In 1836, Graham led a band of about fifty men of various nationalities to help José Castro oust Governor Gutierrez, who was unpopular with the Californios. Graham agreed to help Castro with the understanding that Juan B. Alvarado would be the new governor in California, and that the territory would be independent from Mexico. Alvarado did become governor, California remained part of Mexico, and Graham and his men were rewarded by being imprisoned and sent to Mexico. Upon arrival they were freed, and somehow Graham came out of the deal with $36,000 which he used to gain control of the Zayante tract near present day Felton.[41]

The Californio leaders knew that a dysfunctional Mexican government, which was almost constantly embroiled in its own revolutions, could not help them deal with the Americans. The question for them was: "Under whose umbrella do we want to be?" Some were in favor of coming under the protection of France or England. Others, like General Vallejo, argued that those two could not help, at least not for long, so they should join with the Americans. But true to the then-California way, they did nothing. The fact is, the wealthy Californios wanted self-government. In effect, they nearly had it. They had a good thing going and did not want anyone interfering.

But that could not be. The Americans kept pouring in. In spite of the fact that preparations for a large scale migration had been widely publicized in the US, and that some American newspapers were stating that this was a step toward the inevitable acquisition of California by the US, the immigrants were, nearly without exception, given a kind reception by the Californios.[42] That is, by the Californios—not by the Mexican government. There was a large difference by that time, especially in the north. The news that the new immigrants heard was also quite conflicting, as was the way different individuals handled the news. In 1841, when the Bartleson-Bidwell party arrived in California, about half their number, wanting to conduct themselves lawfully, came to San Jose to request permission to settle in California.[43] Unlike those in their party who stayed behind, these men had confidence in the Mexican government. The local authorities, having received explicit instructions on how to handle newcomers, put them in ward in San Jose. They assured them that this was really just a formality and that some technicalities needed to be worked out. What needed to be worked out was agreement between the Mexican government and the northern Californios, and in fact that wasn't going to happen any time soon. After being detained for about a week, they returned to the rest of their party with a better understanding of, if not a warm appreciation for, the Mexican government.

To say that the Mexican government was unstable is a gross understatement. In the twenty-five years that Mexico ruled California, Mexico had one emperor and thirty-four presidents. Alta California had seventeen civilian governors during that time. Well did Simón Bolívar, the "liberator" of most of Latin America say at his death, "(Latin) America is ungovernable; he who has served the cause of revolution has plowed the sea."

Added to the Mexican government's instability was the fact that communication to and from Mexico took several months. This being the case, it is easy to understand why the Californios were so independent-minded. To make matters even worse, in California the civil and military factions of the Mexican government could not agree. Eventually, in 1844, these disagreements came to a head, and the central government sent General Micheltorena to run the whole show displacing Governor Alvarado and the military governor, General Vallejo. Micheltorena headed to the north of Alta California with 300 soldiers with the intent of diffusing rebellious factions there.[44]

This requires some explanation. Sutter, Vallejo, Graham, and some other powerful men had what were in essence little fiefdoms within Mexican California. They also had their own militias made up of men who benefited by living with them. If it was to their advantage to assist the Mexican government, they did. If the Mexican government crossed them in the least, they were ready to rebel. Sutter, nearly from the beginning of receiving his land grant, threatened to give his allegiance to France. Vallejo, who had rid his land of non-conforming Indians and thus secured a name for himself, was not about to have anyone tell him what to do. And Graham was certainly set on rebellion.

So Micheltorena came north to show Alvarado and Vallejo, and any other rebels, who the boss was.[45] The problem was that his soldiers were actually convicts with military uniforms on.[46] Whether they were all criminals or whether this number included political prisoners is unknown, but when they were not paid, they began plundering the citizenry. The looting and terrorizing of the citizens by these men caused Alvarado and Vallejo to settle their differences, and they, with General Castro, threatened to declare California independent.[47] These events took place nearly two years before the "Bear Flag Revolt" staged by some apparently less than upstanding Americans.[48]

Micheltorena, seeing the locals up in arms, paid 11,000 dollars for passage on the ship *Don Quixote,* and with some of his men high-tailed it back to Mexico.[49] What happened to the rest of the soldiers/convicts is anybody's guess. Most likely many stayed. Add to the northern Alta California cast of characters a heavily armed John Charles Fremont party,[50] sent by Uncle Sam to survey the southern limits of the Oregon Territory, but who apparently had conquest in mind; at least one US naval vessel in San Francisco Bay, and a highwayman named Sanchez, who was robbing and murdering at will in the San Jose area. Throw in the memory of the Texas revolt of a decade before—when that territory was lost to a new republic made up mostly of Americans—and it is easy to understand the uneasiness of the Mexican authorities.

In Texas and in California there were far more Americans than Mexicans who were interested in settling on the land. In both places, the early American settlers received the blessing of the Mexican government. But too many came, muscle began to be exerted, and before long there was a Republic of Texas in 1836, which was followed ten years later by a war between Mexico and Uncle Sam.

The Mexican-American War was declared less than a month after a flag with a Grisly Bear on it was first flown over the town of Sonoma.[51] Acquisition of California by the US was part of the Treaty of Guadalupe Hidalgo on February 2, 1848.[52] Thankfully, in California, the war consisted almost entirely of maneuvering and there was little bloodshed.

The Americans that came to California were of a very different disposition than the Californios. Some of them were arrogant and lawless, some were respectful and law abiding. But whatever they were, they were not laid back. For good or ill, they brought with them their industriousness. They were farmers, tradesmen, and merchants; and upon arrival they immediately started clearing land to plant crops, building homes and what they needed for infrastructure, and setting up businesses. All this was happening before a man hired to build a sawmill found some yellow rocks in a creek in the foothills of the Sierra Nevada Mountains.

There is a fever associated with the yellow rocks and the Incas and the Aztecs had it, as did their conquerors who wanted their yellow rocks. As a matter of fact, perhaps most of the people in the world were affected by the yellow rocks at that time, even though no one had yet found any practical use for them. All they were used for was personal ornamentation, or as an object to boast with in a way that said, "I have something you don't have." So wars were fought over the stuff.

The Ohlone people did not have the yellow rocks, nor did any of the other California natives. Instead, they liked a red rock, which they used to paint their faces and bodies with. The fact that doing so made them sick was a minor inconvenience for something that made them look so attractive. So they fought with neighboring peoples over the red rocks.[53]

Strange folk, these *homo sapiens*.

One of the New Almaden Mines
Photo Courtesy, History San José

5
The Tidal Wave

After the news leaked out, the rush for gold began. But not everyone came to California for the gold. As a matter of fact, after the first year or two, most did not. But the gold sure lubricated all the other enterprises. Many people came, as before the gold rush, to farm, to practice a trade, or to establish businesses; and the northern part of California was flooded. Interest in the land in the Santa Clara Valley increased a thousand fold.

By far the largest problem that Uncle Sam faced after taking over California was determining who owned the land. Most of the land in the Santa Clara Valley had been divided into large ranchos of at least several thousand acres, and the cattle had the run of the place. Where they roamed, they roamed. In Pueblo San José, whether a land title was good was not as big a concern as whatever the existing conditions were. Furthermore, the homes were haphazardly placed and there were no streets laid out. The Americans wanted to see a land title, of which there probably wasn't a "complete" one, by American standards, in all of Alta California. Also, growing crops—what most of the Americans[54] wanted to do—is different than running cattle.

The Mexican Californios surveyed nearly nothing, had no formal system of law, and had an equally informal system of land titles. There had not been the need to do otherwise. But now there was an influx of settlers or squatters—choose your term, in most cases there wasn't a difference—and there was a real mess to straighten out. So Uncle Sam inherited the responsibility of separating the land grants from the land that had now come into the public domain. For this purpose, in 1851, the US Congress passed "*An Act to Ascertain and Settle Private Land Claims in the State of California*." Land that had not been granted by the Mexican government, or was not determined to have proper title, was to be included in the public domain.

The *Act* was to be enforced by a Land Commission with appeals going to the courts. The Land Commission quickly proved useless, as every claim was appealed and went to court.

The Mexicans had no lawyers. What mattered most to them were the traditions of the people.[55] The Americans had lots of lawyers. Still do. And what mattered most to them was precisely how the law was written. Or re-written.

The Americans brought in a virtual army of surveyors in order to make sense of the land claims. The Californios had measured the land from a galloping horse. Literally. Two riders, each holding one end of a rawhide rope of a specified length, galloped from point to point measuring the land while a government clerk watched.[56] No matter that rawhide stretches and the rope was longer at the end of the day. The land was just cattle country, and más o menos (more or less) was good enough.

The boundaries of a few of the Spanish and Mexican land grants in the Santa Clara Valley, as in California as a whole, were at least somewhat clear, but most were not. Many of the grant petitions were filed by illiterate people, and the diseños, or maps, they filed attested to this. To get an idea of what Uncle Sam was dealing with, see the diseños at the end of this chapter. With one look at the diseños, it is easy to understand why most US Senators and Congressmen considered the claims fraudulent. In fact few were. There was just a big difference in culture. [57]

The *Act to Ascertain and Settle Private Land Claims in the State of California* placed the burden of proof of ownership on the landowners—at their expense. This was a reflection of the fact that many in Washington believed that most of the land claims were illegitimate. Other American leaders and citizens wanted to believe they were illegitimate, and still others were determined to legally make them so—even if they weren't. Therefore, the *Act* opened the door for a sea of additional claims. Eventually, there were claims filed for more than 200% of the land in the state as a whole. In the Santa Clara Valley, and in other places where the soil was rich, the percentage was much higher. The average claim took seventeen years to settle, all judgments were appealed, and none of the granted land in the Valley was confirmed by the US Government until 1857. Those landholders who could not afford to file the claim fees and pay the taxes and necessary attorney fees forfeited their claims or sold the rights to them.[58] At least a few attorneys got half of a California rancho for helping secure its title.[59]

One of the largest driving forces behind the land struggle was the American hatred for aristocracy. The concept of huge land holdings of tens of thousands of acres seemed unjust to them. Such a state of affairs had all the charm of the feudal system in Europe from which their forefathers had fled, and they weren't going to put up with it. Another large driving force during this time was the American concept of free land. Free land was the American way from the beginning; it was deeply imbedded in the American mind. People just went out into the frontier and settled. As mentioned in the last chapter, most of the Americans who came to California were from the frontier states. They didn't think California was, or should be, any different than anywhere else they had been. No doubt many had the attitude that if their fathers had driven out Indians before, what were a few Mexicans?

For a Californio to win a land claim in court was one thing. To be able to pay the legal bills and hold onto the land during the long litigation process was quite another. Free range cattle ranching did not mix with the thousands of farmers who were pouring into the Santa Clara Valley. It was virtually eliminated, and with it the means by which the rancheros, or ranch owners, had made their living. The only thing they had left of any value was their land; so shares, or partial interests, in rancho properties were sold throughout this period to come up with the cash to pay the legal fees.[60] Most of the grants in the Santa Clara Valley, as well as in the state as a whole, were eventually confirmed, but in the Valley all the ranchos were lost to legal expenses.[61] Just as the Spanish livestock had gobbled up the livelihood of the Ohlone, so the American legal system gobbled up the land of the Californio Dons.

There were many other factors that contributed to the aptly named "Wild West" of those days. Settlers had seen the value of land skyrocket in the Midwest after it was settled and believed correctly that the same would happen in California. The lure of the quick buck was powerful, and land speculation of every imaginable kind was rampant. Mix together the absence of surveys, the poorly drawn title documents, the phony titles, and a good number of squatters who really believed that they would get a land title just by working the land, and you begin to get a picture of the times.

During this period, "shotgun titles" were common. People squatted on land that seemed vacant enough to them and defended it with arms. There was even a full scale "Settlers War" in the Gilroy area where a thousand squatters took up arms after one of the rancho titles was confirmed by the US Court.[62] A sheriff charged to evict them was unable to get any help to do so. Bloodshed was averted by allowing most of those squatters to purchase the land they had settled on.

In the western hills of the Santa Clara Valley, a man named McCarty settled on a tract of land that he believed was in the public domain. In the early 1850s, he surveyed it, laid out a town, sold lots, and built a toll road into the mountains to William Campbell's saw mill. Later, it was determined that McCartysville was actually in Rancho Quito. Residents, having found the name McCarty less palatable after paying for their lots a second time, eventually changed the name of their town to Saratoga.

The pueblo common lands eventually became part of the city of San Jose. The city, greatly indebted after failing as the state capitol, and with creditors pounding on the door, sold a large tract of pueblo common land at a very unpopular sheriff's auction. A syndicate called The San Jose Land Company purchased it. Though the Land Company presented a huge mound of "evidence" and claimed that the tract was part of a concession of the Spanish Crown, no formal concession or grant could be found. In fact there was none; the Spaniards and Mexicans had not thought it necessary.[63] The sale was resented by many citizens of the city, and the land company became known locally as "The Forty Thieves." Some of the settlers on this tract of land thought they had settled on US government land, or that the land belonged to the individual from whom they had purchased it. They weren't very happy about the sale, and eventually fourteen of these parties filed suit and got a compromise out of the US Supreme Court. The court's decision was good only for those who had enough money to join the suit.[64] The rest moved on.

As the Union was struggling to hold itself together under the cloud of slavery, the laws dealing with California lagged far behind the events taking place in the state and were changed several times. In effect it was mob rule. As more and more Americans came, the laws reflected an increasing anti-Mexican stance. Squatter laws were passed which grew increasingly unjust. At one point, in 1856, an act of Congress declared that all California land was to be placed in the public domain until legal title was shown. It also stated that possession would be evidence of the right of possession. If that wasn't enough, the law stated that a claimant could claim nothing for prior use of the land. And to top it off, the act stated that if an ejectment suit was successful, the plaintiff must pay for any improvements that the squatters had made to his land. The act was overturned by the Supreme Court a year later, but it shows the spirit of the legislation at that time.[65]

The stew of corruption and the resulting confusion affected everyone in California. They, like we, were tried through the events of the times. Some dealt honestly and in good faith. Others were guided by no principle other than the desire for personal gain. The news of the day was rooted in the same soil as it is today—it was biased at best, and often purposefully misleading.

A large number of settlers who came to California in the late forties and early fifties truly believed that there was free land in California, and all they had to do was come and claim it. Many settled on what they thought was, or soon would be, government land. Other squatters killed the rancho cattle, fenced in springs, and raised crops, then used their gains to hire attorneys to wage legal warfare against the landowners.[66] Some even squatted in the mission gardens.[67] Others sold everything they had and came with the intent to purchase land, and did—from someone who did not own it. Some had to purchase the same piece of land two or three times in order to retain it. Still others came with script, that is, official US issue certificates for free land—if they could find any that nobody owned.[68] Some built houses on skids and dragged them from place to place wherever they thought they had the best chance of obtaining ownership of the land. Many acted with honesty and integrity. Others came with the intent to take what they wanted by force. Some changed along the way. Not a few met serious challenges when they realized that they had lost all their savings in one way or another. And so on. As the events unfolded, there were numerous tests on everyone's character.

Many squatters eventually got the land they had settled on; other squatters lost the land they had settled on to other squatters. If the land was determined to be in the public domain, the land was usually obtained for the cost of the legal fees. If it was determined that the land belonged to someone else, in many cases the squatters were allowed to purchase it. In that case, how much they should pay was hotly contested. Raw land was one thing; improved land was another. It took nearly thirty years, but by the mid 1870s there was finally order in California. And very few landowners with Hispanic surnames.

It may be that not one dollar actually went into the US Treasury for California land that was determined to be in the public domain. The historian, Bancroft, states, "In a sense there was no government land to be purchased; every occupant felt that his possession was threatened by squatters on the one hand and grant owners on the other."[69]

Details aside, the US adopted a legal system that allowed California landowners to be plundered by speculating attorneys who argued their cases in front of elected judges. Petty irregularities were allowed to be repeatedly argued in order to bankrupt the land owners.[70] Bancroft summarizes, "As it was, the estates passed for the most part into the hands of speculators who were shrewd enough and rich enough to keep them."[71] The result was ruinous for all—the Californios lost their land, a mockery was made of the rule of law, and the conscience, that guardian of the soul, of many people was seared past all feeling.

What happened to the few remaining Indians in the Santa Clara Valley? First, due to lack of immunity to the white man's diseases, they continued to die off. Second, since those who remained spoke Spanish, most of the Americans simply thought they were Mexicans.[72] By 1900 the Ohlone as a people had disappeared.

The Americans did not respect those who were on the land before them any more than the Spanish and Mexicans did those who were before them. Both groups were fully devoted to their way of life and as a whole saw little, if any, value in their predecessors' ways. A powerful new wave had washed over the land, and other than some place names, it swept away nearly all that had gone before it.

What is man that You should... visit him every morning, and test him every moment?
—Job

Next page:
Diseños, or maps, as filed for Rancho Quito (top) and Rancho Rinconada de Los Gatos land grants

Courtesy of the Bancroft Library; University of California, Berkeley

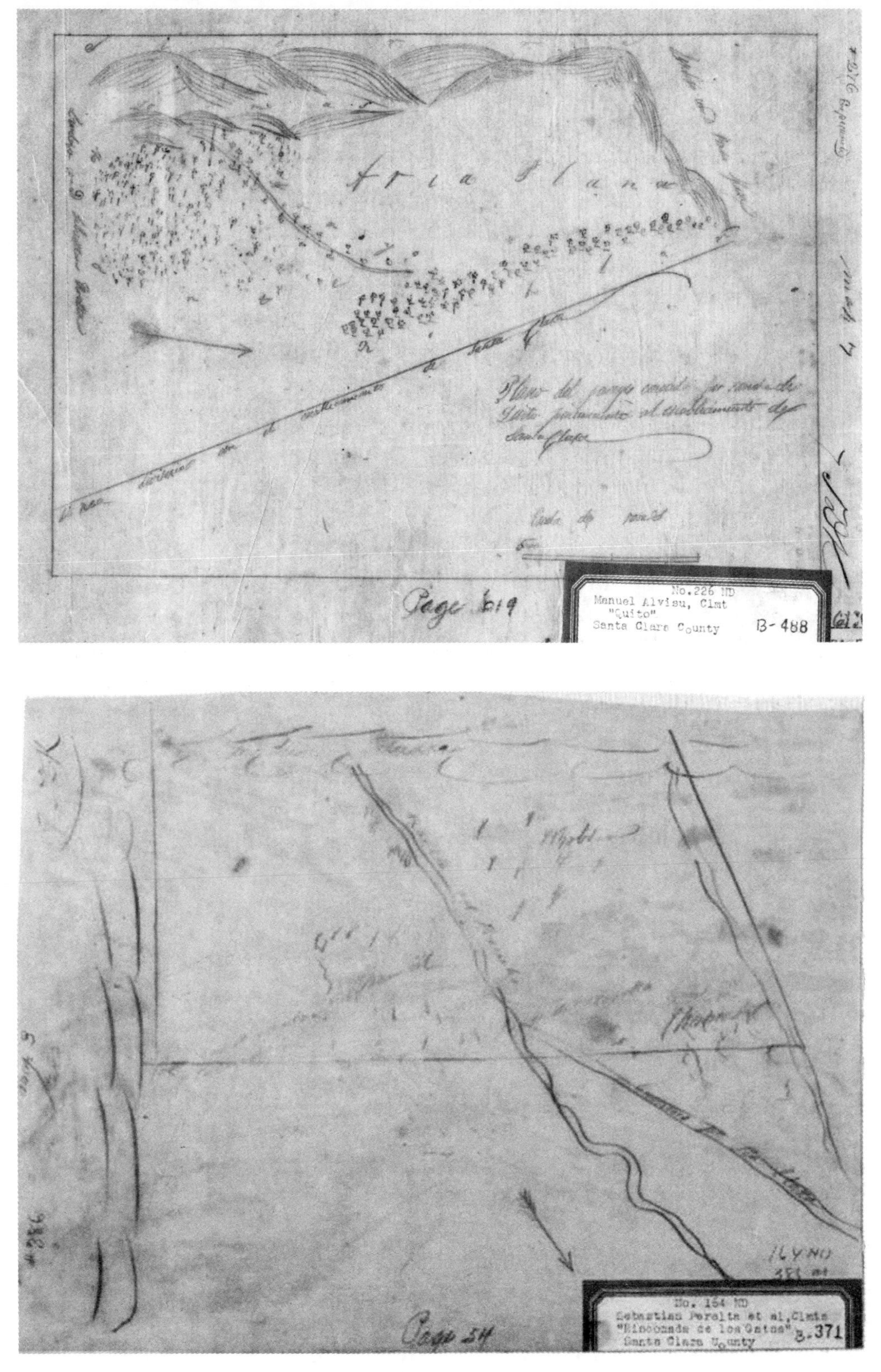
Page 519
No.226 ND
Manuel Alvisu, Clmt
"Quito"
Santa Clara County
B-488
Page 54
No. 154 ND
Sebastian Peralta et al, Clmts
"Rinconada de los Gatos"
Santa Clara County
B.371

6
The Wheat Farms

The American occupation of California was just that. It was a squatter conquest. There were squatters everywhere in the Santa Clara Valley and in the surrounding hills doing all kinds of farming or working in other ways to make a living. The need for food in San Francisco and in the gold country was enormous, and the Valley played a big role in meeting that need. From a few hundred people at the beginning of the American occupation, San Francisco's population had swollen to 100,000 by 1859, and the lack of an established agricultural base manifested itself very clearly in food prices all through the 1850s.

From the early 1850s to the late 1870s, the land in the Santa Clara Valley was used mostly for growing wheat.[73] The free range cattle that had roamed the Valley previously were not compatible with growing crops. Barbed wire was not to arrive until 1874, and since farmers are not fond of having their crops eaten up by cows, cattle grazing was quickly relegated to the hill country. Also, the longhorn cattle raised for hides were not much good for beef, so these herds were replaced with beef cattle.

Through the settlement of the land claims, most of the land in the Santa Clara Valley was eventually divided into quarter sections or parcels close to that size. A "section" of land is 640 acres, so these were approximately 160 acre parcels. The great majority of the parcels became wheat farms.

After nearly seventy-five years of Spanish and Mexican occupation, most of the forest of oak trees that had covered the Valley had been cut down. However, many of the tree stumps remained, and the new mechanical harvesters didn't like tree stumps.[74] Stump pullers were in use everywhere across the Valley, powered by oxen, draft horses, or mules. These devices were essentially wagons which, instead of having a bed, had a beam placed between and high above the axles. Typically the axles were placed about 14' apart. The stump puller, with a chain attached to the beam, was positioned over a stump, the other end of the chain was wrapped around the stump, the animals were urged forward, and the stumps yielded. The stumps were then dragged to the edge of the field and the resulting holes were filled in to level the field.

The wheat fields were dry farmed, that is to say not irrigated, because the winter rains were usually adequate to bring the crop to maturity. After the rains stopped, the crop would dry out and then be ready for harvesting. As late as 1880, from San Jose to Gilroy there were 30 miles of nearly unbroken wheat fields.[75] At that time, California wheat was superior to almost all other, and California soon became the nation's leading wheat producer. The Santa Clara Valley was a big part of the wheat boom in those early years, with wheat production topping out at 1.7 million bushels in 1874. By that time, there were much larger wheat farms in the Central Valley, which in later years took over most of the wheat production.

Although wheat was by far the predominant crop, barley, oats, and rye were also grown in the Santa Clara Valley.[76] Grass seed for livestock forage and hay was also brought in. Many of these grasses look very similar to, or are, primitive grains—wild oats, wild rye, wild rice, etc. The imported grasses did very well on their own, and they pushed out the native grasses. It is the imported grasses that now cover the eastern hills and just about every other bare piece of land in the state where wild grasses grow.

Mills for grinding flour were built on Los Gatos and Saratoga creeks at the base of the western hills. The Los Gatos mill was a failure from the beginning, as the creek did not contain enough water to operate the mill for much of the year. The Saratoga flour mill, known as Bank's Mill, operated until it was destroyed in the great earthquake of 1906. Other mills were built down on the Valley floor, and because there was not nearly enough water power in the area, they were powered by steam. There was one interesting exception: one of the mills was located on top of a powerful artesian well, and was powered by artesian water pressure.

A paper mill, that used straw instead of wood to make pulp, was established in Saratoga in 1868, as was a pasteboard mill that converted straw into a cardboard-like product used for packaging.

Pasteboard was also used in the way gypsum drywall is used today, as an inexpensive interior wall and ceiling covering. Many houses in the area had pasteboard walls and ceilings. The pasteboard mill was located across Lumber Street (today's Big Basin Way) from the paper mill, roughly in the area of what is today Wildwood Park.

The farming of grains requires either crop rotation or occasional fallow years to replenish the soil.[77] Much of the Santa Clara Valley soil was so rich that it virtually could not be depleted. But in some places, because of poor farming practice, yields dropped dramatically after a few years, so orchards and row crops, such as peas, beans, and berries eventually replaced wheat.

7
Red Rocks, Lumber, Oil and Straw

Besides the need for food, the need for machinery and other manufactured goods was very great for the new empire being built in California. There was little time to get needed goods elsewhere, so as much as could be found or made locally, was. A gold rush doesn't wait, and the mining industry's need for supporting industries was enormous. San Francisco was the hub of the empire, and the Santa Clara Valley was close to San Francisco and supplied both the city and the gold country. This chapter focuses on three major non-farm industries of the Valley and surrounding area during the early days of the American era. The red rocks that the natives had coveted brought forth an extract that was being produced in huge quantities; the redwood trees in the coastal hills yielded the lumber that built San Francisco; and the fuel needed to drive all the new industries was obtained in several ways.

Red Rocks

In 1845, a Mexican officer named Andres Castillero went out to the southern part of the Valley to see the red rocks that the Indians were so enthralled with. He was trained in mineralogy, recognized the red rocks as cinnabar, and filed a claim for the Indians' mine with the Mexican government. The area was eventually named New Almaden, after Almaden, Spain. Up until this new discovery, Almaden, Spain had been the leading producer of mercury in the world.

To extract mercury from cinnabar ore, the ore is crushed, then cooked in order to release mercury vapor. The vapor condenses into a liquid when it comes in contact with a cooler surface, just like water droplets condense on the outside of a glass of ice water. In the earliest days, the ore was cooked in huge pots the miners had obtained from a whaling ship. Later, increasingly efficient means were used. As the value of the mines at New Almaden became apparent, there were, of course, disputes over the title to them, and eventually legal wrangling caused a shutdown of the mines for more than two years. In 1864, after the lawyers had consumed the money of the smaller players, the operation was sold and re-opened as "The

Quicksilver Mining Company." The company had big money behind it and went to work cooking ore. Lots of ore. The quality of cinnabar found in New Almaden was such that twenty-four pounds of liquid mercury could be extracted from a ton of ore. More than 26,000 *tons* of liquid mercury was extracted from the New Almaden mines by 1880, and by 1887 half of the world's supply of "quicksilver" came from California, with the greater portion of it from New Almaden.[78]

This mercury was not going into thermometers. It was headed to the gold country where it was used to separate gold from gold-bearing gravel or crushed ore. The use of mercury for gold separation has been used since Roman times, if not before. In California, most of the mercury produced was used in hydraulic mining. Huge water cannons broke down gravely hillsides, and the resulting slurry was directed into long troughs, called sluices, with ridges built into the bottom. Liquid mercury was poured on top of the gravely mixture as it went down the sluice. The density of mercury is close to that of gold, so the mercury would stick to the gold and the amalgam formed of the two would be stopped by the ridges at the bottom of the sluice as the sand and gravel passed over the top. Unfortunately, about ten to thirty percent of the mercury was lost in the process. In some cases the mercury was simply burned off. Mercury is not a healthy substance, and though delayed, we are paying the price for that practice today.[79]

Lumber

Another early Santa Clara Valley industry, probably as large economically as the wheat farms and the quicksilver mines, was lumber. The Santa Cruz Mountains on the coastal side of the Valley were, and thankfully still are, covered with coastal redwoods. Lumber from these stately trees is straight, strong, nearly clear of knots, and resists decay. In other words, it is perfect for building. Other commercial lumber species in the area included Ponderosa Pine, Douglas Fir, and Black Oak, but the harvesting of redwood dwarfed that of all other species combined. Redwood was used far more than any other wood species for the building of San Francisco, San Jose, and the other towns and cities around the bay. Wharfs on the bay were also built of redwood, and later the great majority of railroad ties and bridges in the state were built from it. So great was the demand that many miners quit the mines to work in the mountains harvesting lumber. During the boom years—which lasted until the Civil War—a man could make $16.00 a day in some places, which was a huge salary in those days and equal in value to an ounce of gold.

There were two main routes to bring the lumber down from the mountains into the Santa Clara Valley: Los Gatos Canyon and the Saratoga Gap. Lumber cart traffic was heavy and the dust was thick on the roads coming down from the mountains, every day except Sunday. The lumber that was not used in San Jose was shipped out from the shallow water Port of Alviso that is on the southern tip of the bay.

Alviso had been the loading place for the cow hides during the Spanish and Mexican eras, and with the gold rush on, barges loaded with nearly everything went in and out of the port. Passenger steamers, which were the preferred mode of transportation around the bay, also made regular runs from 'Frisco to San Jose. At one time, it was thought that Alviso would become a great city. That was before the railroad came to San Jose in the mid 1860s, after which time Alviso was turned into a virtual ghost town. [80]

In the late 1870's, a railroad was built through Los Gatos Canyon that connected some of the richest timber area to the Valley. By the early 1880s, with the aid of four tunnels and dozens of bridges, it had reached Santa Cruz. Those hills are probably more pleasant in appearance today than when they were all logged off.

Coal and Oil

Fuel for driving the new industries was in short supply in the early decades of the American era. The largest local coal deposit was found at Mount Diablo about thirty miles northeast of the Valley. There was even a vein of coal struck while digging one of the tunnels mentioned above, but the demand far exceeded the local supply. Both San Francisco and Sacramento had many foundries, and most of the coal to fire them was shipped in from Australia. San Jose's first natural gas, around 1860,[81] was extracted from Australian coal. But this was expensive energy and there was also another way to get a hot burning fuel.

Remember all those large oak tree stumps mentioned in the last chapter? In the right hands they, and waste wood from the logging industry, made good charcoal.

A charcoal oven was easier to build in the hills than in the valley.[82] In the flats, a pit needed to be dug; but in the hills, a suitable small ravine was found, and most of that work was already done. An area was cleared, and large stumps, dragged by draft animals, were put in first. Other wood was piled on top in decreasing size, and then the whole pile was covered with a thatching of branches. Next, the thatching was covered with earth, except for a chimney hole in the top of the pile and a narrow opening at the bottom of one side that served as the oven door. A fire was started at the oven door and was monitored carefully. Controlling the air controlled the fire. Additional air from bellows pumped into the oven door would make the fire hotter; closing the chimney opening a little by adding more dirt would slow down the burning. If too hot, instead of producing charcoal, the wood was burned up; if not hot enough, the fire went out; if just right, in two weeks a good pile of charcoal had been produced. All that remained was to dig it out, put it in sacks, and cart it into town… And you think your job is tough?

Obviously there was the need for a better, cheaper fuel source. The answer was found in Los Gatos Canyon: oil.[83] In the fall of 1861, oil was discovered in the hills near Lexington, and in the mountain area from Alma to Wrights, oil fever took over.[84] A good number of companies got involved, a lot of stock was sold, and several oil wells were dug. The wells averaged about a thousand feet deep and the best well produced about a hundred barrels a day. It was the most productive oil well in California at the time. The premiere oil area was in Moody Gulch, and a three mile pipeline brought the oil down from there to near the edge of the flats where it was stored in a large tank. Before the railroad came, the oil was put into barrels, loaded onto wagons, and carted into San Jose. After these and other oil finds in the surrounding area, it looked like the Santa Clara Valley area would be the oil producing region of the state. Standard Oil even built a refinery in San Jose.[85] There was plenty of excitement about the discovery, but like most other discoveries, the find was exaggerated. It was eventually determined that there was not nearly the "inexhaustible supply" that was first reported. Then, oil was discovered in Southern California in quantities that dwarfed the finds in and around the Santa Clara Valley, and thus ended the Valley's oil boom. But there was enough oil found that, when refined into kerosene, supplied lighting for the area. Tar for roofing materials, and lubricating oil for machinery, were also early oil-based products produced locally. Refining methods for making fuel for heating and running machinery were not perfected until the early 1880s. The canning industry boomed at that time largely due to this development.

Straw

In those days, before the internal combustion engine, power was also needed in the wheat fields. But how to get it?

Enter Joseph Enright.[86] His invention, the Enright Portable Straw Burning Engine, was a steam engine on wheels that was designed to burn straw. With nearly the entire Valley covered with wheat and other grains,[87] there was no lack of straw for fuel. The steam engines revolutionized wheat farming. Threshing and sacking of grain could now be done very efficiently in the field. Excess straw that was always a nuisance was put to good use, and due to the efficiency of the operation, the cost of wheat dropped dramatically. At one time there were over two hundred of these engines used in California. (The down side to Enright's engines I can't help wondering about: they were straw burning engines operating in the middle of dry wheat fields… Anyone else see a problem with that? Seems to me that Enright, for all his genius, deserves a chapter in *The Adventures of Dangerous Dan.*)

This is just a sampling of the early industries in the Santa Clara Valley at the beginning of the American period that were closely tied to the land. There were many more. For better or for worse—and there was certainly both—in less than a decade the Valley was transformed into part of the American industrial machine with all the activity of a beehive on a summer's day.

Next pages: Enright's engine, at left, driving a threshing machine at right. Note the large drive belt connecting the machines. Cut wheat (center) is being moved to the thresher by use of a tripod derrick. Sacks of grain are at the far right. At far left, straw is being fed into the engine. Photo Courtesy, History San José.

8
From Rancho to Farms

This chapter will go back a little in our story to take a closer look at one of the Santa Clara Valley ranchos. It will tell what is known of that rancho, of its breakup, and of some of the people involved. Doing so will put a human face on the transformation of the rancho into many farms. The story recorded here is typical of how that transformation took place on all of the ranchos in the Santa Clara Valley.

On the western side of the Valley, and extending into the Saratoga hills on the south, was Rancho Quito, a Mexican Land Grant of about 13,152 acres. Through the middle of the rancho ran Arroyo Quito, which is now called Saratoga Creek. The rancho included most of what is now the city of Saratoga and extended north to include parts of today's Campbell, Cupertino, San Jose, Sunnyvale, and Santa Clara. It was roughly joined by Rancho Rinconada de Los Gatos on the southeast, and Mission Santa Clara property on the northeast. Otherwise it was surrounded by unclaimed land.[88]

Rancho Quito was granted to José Noriega and his father-in law, José Zenon Fernandez, in 1841 by Governor Alvarado. Both of these men had been members of the thwarted Hijar-Padres colonial expedition.[89] Fernandez became Pueblo San Jose's first paid school teacher and later held other civic offices in San Jose and at the capitol, Monterey, during the 1830s and 40s.[90] Noriega also held various government positions, and in addition to the Quito grant, received large land grants in what are today Alameda and Contra Costa counties.[91] In 1844, ownership of Rancho Quito went into the hands of Ignacio Alviso, who had been granted Rancho Rincon de los Esteros six years earlier.[92] Alviso had come to California with the Anza expedition at age two with his father, Domingo, who was a soldier.[93] After he became of age, Alviso was one of the foremen at Mission Santa Clara who directed Indian labor.[94] He is written of as a man of intelligence and extraordinary talent and is credited with overseeing much of the building at the mission. Later, he served as administrator for the mission. Alviso and his family lived in the mission compound as did the priests, the Indians and the guards. The town of Alviso, named after Ignacio, is now part of San Jose. It was established on part of Rancho Rincon de los Esteros as a shallow water port on the southern tip of San Francisco Bay.

When Ignacio Alviso died, in 1848, Rancho Quito passed to his son, Manuel.[95] This transfer of ownership was litigated for many years by other members of the Alviso family.[96]

Though settling on the land was technically a requirement to secure title to it, it does not appear that Noriega, Fernandez, or any of the Alvisos ever settled on Rancho Quito, though they almost certainly had cattle there. Earlier, at the time the Mission Santa Clara property was taken over by the Mexican government, Ignacio Alviso was put in charge of 3,700 head of cattle that had belonged to the mission.[97] Since the ranch land adjoined mission land, the former mission cows likely grazed in the same area and had the same vaqueros, or cowboys, tending to them. The former mission Indians who stayed in the area had to join themselves to one of the rancheros to survive, so for them the change was not all that great either.

In those days, many of the rancheros in the Valley lived in the pueblo at San José.[98] Notable exceptions were the owners of the neighboring rancho, the 6,631 acre Rancho Rinconada de Los Gatos. José Maria Hernandez and Sebastian Fabian Peralta built a homestead and settled on that rancho, southeast of Rancho Quito. Their adobe home was located near today's Vasona Park.

In 1861, the grant for Rancho Quito was confirmed by US patent to Manuel Alviso, José Zenon Fernandez, and several members of the Fernandez family. But as with all the other ranchos in the Valley, the Rancho Quito property was hotly contested, and by that time most of the rancho land had already been lost to the cost of defending it.

There is an historical record of some of the people who acquired parts of Rancho Quito land early in the American period. This record is largely from a few local history books written in the 1870s and 1880s that give considerable space for biographical sketches of some of the early American settlers in the Santa Clara Valley.[99]

Included below are some biographies found in these sources that apply to Rancho Quito. They give a picture of the background and circumstances not only of these people, but also of the others who came to farm in the Valley early in the American era. The information from these sources was taken at face-value and the verbs describing these settlers' acquisition of their land have been retained. The properties of all the people listed here can be found on the farm map at the end of this chapter.

Reuben and Ellen McCoy. Mr. McCoy was born in Tennessee in 1825. His parents moved to Jackson County, Missouri when he was eleven, and a short time later the family moved to Platte County. In 1850, Mr. McCoy came to California with his uncle by wagon train. He spent about fourteen months mining in the gold country at Placerville, then came to the Santa Clara Valley in the winter of 1851-1852 and settled on a farm of 305 acres where he grew wheat and barley. His farm was located between what is today Quito Road and San Tomas Aquino Road and was bounded on the south by San Tomas Aquino Creek. Mr. McCoy was married in 1866 to Ellen D. England. They had five children.[100]

Samuel A. and Serena Blythe. Mr. Blythe was born in Shelby County Tennessee in 1826. In 1831, his parents moved to Arkansas, in 1834 to Memphis, Tennessee, and in 1837 to Texas. In 1849, Mr. Blythe came overland to California. He came directly to the Santa Clara Valley and worked in the lumber industry. In 1853, he purchased a 200 acre farm three miles north of the village of Saratoga near the current intersection of Quito Road and Saratoga Avenue on the west side of what is now Quito Road. The Blythes married in 1852 and had seven children.[101]

William and Dicey Cox. Mr. Cox was a brother-in-law of Samuel Blythe. Mr. Cox was born in Ohio in 1827. His father was from Virginia and went to Ohio when he was eight years old. William Cox was married in 1848 to Dicey Baggs, another Ohioan. Mr. Cox, his sister, Serena (later, Mrs. Blythe), and his father, John Cox came by wagon train to California in 1852. The trip took six months. Upon arrival, William hired out as a farm hand, and the next season he rented a piece of land from Samuel Blythe. In 1854, the Cox family bought seventy acres adjoining Mr. Blythe's farm to the southwest. A few years afterward, they bought more land and eventually had a farm of 312 acres where they grew hay and wheat. The Cox farm was along what is today Cox Avenue in Saratoga. William and Dicey Cox had seven surviving children.[102]

Ira J. and Ann L. Lovell. Mr. Lovell was born in Logan County, Kentucky in 1811; Ann (L. Campbell) was from Muhlenburgh County, Kentucky. Mr. Lovell's parents moved to Muhlenburgh County shortly after his birth. In 1839, after the Lovells married, they moved to Hopkins County and farmed there until 1850 when they moved to Saline County, Missouri. In 1852, the Lovells, with seven children at the time, headed across the plains to California. In the fall of 1853, they homesteaded on their farm of 235 acres. The Lovell farm was between San Tomas Aquino Road on the west and San Tomas Aquino Creek on the east and south. The Lovells had eleven children.[103]

James M. and Martha Kenyon. Mr. Kenyon was born in Ohio in 1817 and was trained as a carpenter. After he became of age, he moved to Missouri and married Martha Roberts there. They came overland to California in 1849 and lived in Coloma, in the gold country, until mid-1850. In the fall of 1850, the Kenyons moved to San Jose and built a house there. Shortly afterward, they purchased their 240 acre farm about two miles west of downtown Santa Clara where they grew hay and grain. By the 1880s they had begun planting fruit trees. The farm was just east of today's Lawrence Expressway at Pruneridge Ave. The Kenyons had five children.[104]

These Americans, who moved onto Rancho Quito land in the early 1850s, could be readily traced. All remained on the land at least until 1880. Their backgrounds are similar and are typical of those who came to the Santa Clara Valley to farm at that time. During the long litigation of the Rancho Quito claim, there were likely hundreds of others who settled or squatted on the rancho and who bought and sold parts of it. There are many other interesting stories of the early settlers besides these and the previously told story of McCartysville.[105] One of particular note is that of the Arguello family.

In 1859, while the claim for the rancho was being contested, José Ramón Arguello, the son of Luis Arguello, the first Mexican governor of Alta California; his mother, Maria Soledad Ortega de Arguello; and a business partner named S. M. Mezes obtained control of a large part of Rancho Quito. Unlike Noriega, Fernandez, and Alviso, Arguello did settle on the land, and in the early 1860s he planted an olive orchard near the current intersection of Quito Road and Saratoga Ave. Mezes seems to have disappeared from historical record. [106]

As had happened to Manuel Alviso, the cost of defending the rancho land from the various claims on it forced the Arguello family to sell off much of the remaining land, and by 1876 they owned a mere 631 acres—less than five percent of the original rancho. José Ramón died in 1876, and after a few years the Arguello heirs sold out to a man named Goodrich. Thus ended one of the last large land holdings of the Californios in the Santa Clara Valley.

Next page: Part of the Thompson and West Atlas Map of 1876 *that shows the division of Rancho Quito into farms at that time.* [107] *Within a decade after this map was made, most of the larger farms were divided up into the smaller orchard farms that the Santa Clara Valley later became famous for. The beginning of the change from wheat to orchards had already begun by the time this map was drawn. Note that some of the orchards are drawn in.*

The boundaries of the Rancho Quito, as per the current Thomas Brothers Map, are the darkened line and the San Tomas Aquino Creek. The southern boundary extended off the map and south of Saratoga-Los Gatos Rd., which is near the bottom left corner of the map. Rancho Rinconada de Los Gatos is to the east at the southern portion of the map.

Mountain View Road (top left) is Saratoga-Sunnyvale Road
San Francisco Road (top) is El Camino Real
Campbell's Creek (at center) is Saratoga Creek
"Redwood" was a township name
"Moreland" was a school district

SAN FRANCISCO ROAD
BOYTER RD
MILLIKEN
MILLIKIN SCHOOL
YOUNG ROAD
Young Road
Collins School
SANTA CLARA HOMESTEAD ASSOCIATION
BENNETT TRACT
ODD FELLOWS SAVINGS BANK
SARATOGA ALVISO ROAD
STEVEN'S CREEK ROAD
SANTA CLAR
Campbell's Creek
Arroyo De Los Calabazas
Mountain View Road
SOUTH METHODIST
MORELAND SCHOOL
MORELAND
Lincoln School
BUBB ROAD
QUITO RANCHO
SARATOGA AVENUE
SAN TOMAS AQUINO ROAD
QUITO FARM
RED WOOD
QUITO ROAD
San Tomas Aquino
SAN TOMAS
ARATOGA
SARATOGA
RINCONADA DE LOS GATOS
GA
le 40 Chains to One Inch.
Thos Boyter 168.31
I.N. Thompson 127 1/2
W.H. Hall 253.78
W. Tregouth 158
D. Hayes 164
L.B. Clark 80
I.N. Thompson 355.03
L.E. Gardner 164.52
M. Glendenning 195.35
I.N. Thompson 323.19
J.M. Kenyon 242.08
A. Stephan 179.26
Michael Wolfe 162.45
Michael Scanor 146.52
B. Craft
G. Theuerkauf
P.H. Doyle 242
Adam J. Bollinger 160
W.H. Warburton 283
M. Tantau 464
G.B. Walters 125.50
S. Graves 164.60
C.H. Lapham 128
C. Bollinger 279.85
Peter Hillebrant 304.92
S. Graves 190
D. & T. William 356.79
Peter Ball 104.30
Jacob Graves 320
Vine Yard
James Payne 126 1/2
Thomas Scully 360
J.R. Argüello
J. Messersmith 290.60
S.A. Blythe 210
William Cox 312
J.A. Lovell 216.50
R. McCoy 303
J.S. Scott 117.60
T. Boyter 193
E.N. Parr 444.23
H.J. Lovell 237
T. Nolan
F. Laird Est 160.25
W.J. Parr 423.08
C.H. Parr 441.63
T. Kirwin 320 a.
J.P. Bubb 95.43 a.
J. Hickey
W. Horrison 157 a.

A typical tank house. Every farm had one, and several farmers were still using windmills to pump water late in the agricultural period.[108] *Tank houses were built in three levels. The upper level was for the water tank, which unlike this one, was often enclosed. The middle level was used for cool storage, and the ground level was often a shower room or additional storage. The lower floor walls were angled out like a skirt for increased stability.*

9
The Garden of the World

Before the coming of the Americans, an early Californio would never have thought that the Santa Clara Valley could ever be referred to as "The Garden of the World." Rainfall in most of the Valley is less than twenty inches per year, and it usually only occurs between October and April. The best of the streams are seasonal, and for half the year the place is naturally quite dry. It was just cattle country—on the surface.

But there was water, plenty of good, sweet water deep underground. When the Americans came, they began drilling wells. It is not that the Californios did not dig wells; they did. But the Americans went deeper and were handsomely rewarded for doing so. It is one thing to capture ground water by digging a shallow well a little deeper than the water table, the depth at which the soil is saturated. It is quite another thing to drill into a deep aquifer, a gravely underground reservoir. Ground water is quite limited in most places in California due to the lack of rainfall during much of the year. In addition, it usually contains a high concentration of organic matter that negatively affects its taste. Deep well water is entirely another matter. It is filtered far better and tends to be sweet, and if the aquifer is large, it is plentiful.

It turned out that water could be found just about anywhere under the Valley floor and surrounding hills. There was so much water that when the first deep wells were dug, there was often sufficient pressure to bring artesian water to the surface.[109] When this abundance of water was discovered, it changed everything. The possibilities for farming became nearly unlimited.

With plentiful water and the land claims being settled, many farmers were no longer concerned about making the investment required to plant an orchard. Mission Santa Clara had a wonderful orchard that proved conclusively that a large variety of fruit trees could be grown in the Valley. If a farmer could get his fresh fruit crop to San Francisco in good condition—and that was a big "if"—the rewards could be enormous. So, many farmers put in orchards.

From the time of the founding of Mission Santa Clara, newcomers had brought their favorite fruit trees with them to the Valley. Later, from the beginning of the American era, businessmen traveled the world to find what fruit trees they thought might grow well in the area. Honey bees, which are necessary for the pollination of most fruit trees and vegetable crops, as well as for alfalfa and most clovers, were imported from Europe. A man named Christopher Shelton brought the first honey bee colonies to the San Jose area in 1853, and in the 1860s and 70s, thousands of colonies were brought across the plains and across the Isthmus of Panama into California. [110]

Planting an orchard is a *very* substantial investment. A careful analysis must be made of exactly what to plant and exactly where to plant it. Fruit trees must be selected for fruits that will have a ready and constant market. After a variety has been selected, a determination needs to be made if it is best to plant seedlings or cuttings. If it is determined that cuttings are preferable, grafting the cuttings onto the rootstock of another tree—usually a native—needs to be considered. The ground must be prepared adequately. Rocks and other obstructions must be removed. After planting, the trees must be nurtured carefully for about five to seven years before they are able to produce a crop that is worthy of taking up the ground. And labor must be available to bring in the crop.

Perhaps nothing in farming is more risky than putting in an orchard. One only needs to drive around the state a little to see dozens of orchards that have failed. Not enough water, wrong soil type, incorrect climate, a change in consumer tastes, inadequate transportation, labor problems, or any one of a host of other factors can easily bring a lot of hard work to nothing. In addition, planting an orchard is a marriage. Once an orchard is in, it must be tended to or its value will be lost.

The many factors to consider when making the decision to plant an orchard apply equally to far more than just orchards. If a farmer puts in an apple orchard, for example, and it produces two tons of apples per acre, he might be quite happy. He may have a ready market for his apples and may be able to make a good living off of them. But if someone in a more suitable area puts in apples and gets ten tons to an acre, the first farmer had better find something else to grow—and quickly. In the Valley, as everywhere, people were, and are, watching to see what is successful and what is not. What is successful will be copied, the market will become flooded, and the farmer—if he is going to continue farming—must grow what grows best in his area. It is also prudent to spread the risk over a few crops. Competition is a fact of life. Even wildlife competes for food. Grasses compete for space, and it's hard to think of pretty wild flowers or cute deer competing for anything, but they do.

In the early days, while San Francisco and the gold fields were booming, nursery stock, like all other goods, was sold at enormous mark-up. Nevertheless, the demand was very high and many farmers took the risk, bought expensive stock, and put in orchards. By the late 1860s, however, there was a glut of fresh fruit in the area and no market for it, so a lot of it rotted. Disgusted farmers tore out some of those early orchards and switched to other crops.[111] So, the orchard industry in the Santa Clara Valley had a shaky start until the middle 1870s when canning and drying methods were perfected.

Several varieties of many kinds of fruits were tried in the Valley. In the end, the most successful commercial orchard crops, in order of tonnage produced were: prunes, apricots, peaches, pears, walnuts and cherries. A variety of grapes for wine making were also very successful, and viticulture will be touched upon in another chapter. But there is one wine maker, or vintner, of particular relevance that needs to be mentioned here.

In the mid 1850s, a French vintner named Pierre Pellier brought several hundred cuttings from choice vines in the old country to the Santa Clara Valley. He also brought two trunks full of French Prune cuttings for his brother, Louis, a nurseryman. The cut ends of the cuttings, or scions, were stuck into raw potatoes and the trunks were filled with moistened sawdust to keep them from drying out during the long voyage.[112] Louis took the cuttings and grafted them onto domestic plum stock.[113] Little did he know that by 1919 there would be over seven million French Prune trees in the Valley.[114] It turned out that the soil and climate of the Santa Clara Valley made for the perfect ripening of the fruit. Mr. Pellier also brought several other fruit tree cuttings—cherries, pears, plums, and others, but none that he brought were even remotely successful compared to the French Prune.

The French Prune did not originate in France, but in the Middle East. European horticulturists got hold of it, tweaked it a little,[115] and in the true spirit of European nationalism, named their new varieties after their homelands. So there are Italian prunes, German prunes, and French prunes, which are all very similar—at least to the novice. But early on, the French Prune became the standard for the industry, and remained so for the entire fruit-growing era in the Santa Clara Valley.

What made the prune industry, and later the apricot industry, in the Valley, was the ability to efficiently dry these fruits. Fruit had been dried for millennia, but not with the success that was achieved in the Santa Clara Valley. During the late 1870s, the process of preserving prunes by dipping them in hot lye water was perfected. Later, thanks to Henry Coe and others, the sulfur smoking of apricots and pears was also perfected.[116]

To come up with the cash needed to plant orchards, most of the wheat farmers began selling off parcels of their quarter sections. Fruit farming requires far less acreage than wheat farming and is correspondingly more labor intensive. By 1890, most of the Valley land was divided into small family farms, and as a whole they did marvelously.[117] Less some other problems, which will be discussed in the next chapter, with ten to twenty acres of fruit trees a farmer could earn a decent living.[118]

The railroad companies had much to gain by the establishment of these small fruit farms. More farms producing more goods meant more customers and more freight to move. Even before the Transcontinental Railroad was completed in 1869, canneries had begun springing up in the Valley. These were mostly "Mom and Pop" operations—several literally started out using the kitchen stoves in their homes. During the middle and late 1870s, technological advances in sugar refining, steel can manufacturing, and pressure cooker development all combined to transform the

Valley's wheat farms into orchards.[119] The boom was so fast that by 1880, hundreds of tons of canned fruits were being shipped out of the Valley, mostly by large canneries. By the end of the era, the huge food conglomerates had taken the place of nearly all of the independent canneries. Fruits and vegetables for canning were also trucked into the Santa Clara Valley for canning from the Central Valley, the Salinas Valley, and other areas.

After the grain crops had phased out in the Santa Clara Valley and the orchards had been established, there still remained other kinds of farming. Vegetable row crops were grown in the lower lying areas where the water table was high. Sweet peas were a big crop, as were strawberries. But by the late 1880s, fruit growing dominated the Valley.

By 1919, the Santa Clara Valley was covered with 10 million fruit trees, about three quarters of which were prunes. The "Prune Center of America" had come of age. At least forty canneries were operating in the Valley at that time. One of them, the Co-operative plant, was the largest cannery in the world. Total annual production in 1921 from several thousand farms was 100,000 tons of canned fruits and vegetables, 65,000 tons of dried fruit, and 12,000 tons of fresh fruit. About two thirds of the output was shipped to the domestic market.[120] At the peak of the agricultural era, around 1950-1952, this production was tripled. San Jose, in particular, was the thriving industrial center for processing these fruit and vegetable crops.

Valley of Heart's Delight was the brand name of one of the local canneries, and early on that name stuck to the Santa Clara Valley. The Valley was in good measure, "The Garden of the World."

Next page: The Santa Clara Valley in bloom
Photo Courtesy, History San José

10
Orcharding

As any abandoned fruit orchard will testify, fruit trees don't just produce well by themselves. In the spring, the top foot or so of the soil needs to be turned over to cover the ground vegetation that has grown up after the winter rains. This replenishes the soil and aerates it so that it will receive the rain that comes the following winter. It must be done after the rains but before the soil has dried out and become hard. Turning over the earth is called "disking" and is accomplished by using a tractor to drag a trailer-like platform outfitted with several rows of sharp steel disks similar in shape to cymbals. The disks are placed at a very slight angle from the pulling direction so that the earth is turned over ever so neatly covering all the vegetation. (See photo at the end of the chapter.)

However necessary the disking process is, and however wonderful the scent of an orchard being disked may be, it is at the same time a little sad. The beautiful blanket of mustard that grows in the orchards, in all its glorious yellow bloom, is plowed under.[121] But woe to the farmer who waits too long! He will have a very bumpy, dusty ride and may have to run the disk over the ground three times and still not get deep enough. Like everything else in farming, disking must be done at the right time. If the farmer goes out too early, he gets bogged down in the wet soil, and if he goes out too late, he will sure wish he hadn't.

Irrigation of the orchards was interesting. Some orchards, especially in the western hills, needed no irrigation at all. The water table was relatively high in many places, and the cooler climate was also a factor. When the water table is lower, irrigation is often required, especially down in the flat land of the Valley where it is hotter. A down-side to irrigation (other than drip irrigation which did not exist during the fruit growing era in the Valley) is that the bugs love it, so spraying may be thought necessary. Conditions have to be monitored carefully and the irrigation and the spraying done with an adequate understanding of short and long term effects.

Another down side to irrigation, and a far more subtle one, is that it can give a false sense of security. In the last chapter, it was noted that with the discovery of abundant underground water, the farming possibilities in the Santa Clara Valley became nearly unlimited. Nearly. But not unlimited. As the farming industry exploded in the Valley, eventually more water was being taken out of the aquifer than could possibly be replenished. Naturally, the water table dropped. So more irrigation was needed. With excessive pumping, the aquifer began to collapse and when that happens, even if abundant rain comes, it can no longer hold the water it once did. So the land settled, and in some places, such as Alviso, the Valley has settled more than ten feet since the beginning of the American era.

Cherries

The cherry trees blossomed first, so the cherry growers had an additional concern not shared by most of the fruit growers: frost. Their entire crop could be lost in one night. So the cherry farmers had to keep very alert and if necessary, set out smudge pots to raise the temperature in their orchards slightly.

Smudge pots were orchard heaters and are a big part of California agricultural history. They were basically just large oil cans that were filled with fuel oil and outfitted with wicks and chimney stacks. Smudge pots were not pretty, did not smell good, and if it got really cold, could make the air sickening. But until the development of orchard fans, they saved many crops that otherwise would have been lost to frost.

The cherry varieties grown in the Santa Clara Valley were Bings and Royal Annes, and they ripened in late May. Cherry season was announced in many of the orchards by the boom of mechanical cannons being fired every few minutes from sun up until sun down. The booming of the cannons kept the birds, who love the cherries as much as people do, from damaging too much of the crop. When tin foil became available, cherry growers would also put strips of it in the trees. The sparkling reflection would scare off some of the birds, but no matter what was done, some of the crop was shared.

Apricots

Apricots, almost all Blenheims, were next and usually started in late June, right when the kids got out of school. The harvest, of course, is the busiest time of the year and when the largest crew is needed. In the Valley that meant the school kids. For most of the apricot growers, it was not just a matter of picking the fruit and taking it to the canneries. A huge part of the fruit industry was preparing dried fruits. These were mostly apricots and prunes. Due to the abundance of sunshine and nearly unheard of summer rain, many of the farmers did their own drying.

The great majority of apricots grown in the Santa Clara Valley were grown for drying. The apricots were sliced in half and laid sliced side up on wooden trays, which were then stacked onto carts, and then rolled into a kiln where they were smoked with sulfur. The cart wheels and tracks were identical to those of a railroad, but were of a much smaller scale. The trays were about three feet wide and eight feet long, and the kilns were built just wide enough to comfortably accommodate the carts. Apricot kilns were made of just about anything. They were made of steel, brick, concrete, wood, and some farmers simply had a fire pit in the ground, rolled their cart filled with apricots over it, and threw a tarp over it. A sulfur fire was started in the kiln (a ghastly job), the cart was rolled in, and the steel doors were closed. Within two to four hours the apricots were coated with sulfur smoke. Smoking kept the fruit from turning black or having mold grow on it. Also, after smoking, the flies were not interested in it. Unless the sulfur fire was allowed to burn itself out and the smoke to dissipate overnight, opening the doors of the kiln was not a pleasant experience. After they were taken out of the kiln, the trays were then spread out on the ground in a drying area for a few days. The beautiful picture of orange hillsides full of apricots drying in the sun is forever etched in the mind of anyone who saw it. It was similar to, yet surpassed by, the beauty of a thick bloom of poppies in some nearby places a couple of months earlier.

Apricots are a little finicky. They are susceptible to brown rot, which can be brought on simply by a change in the spring weather.[122] Brown rot is sometimes mild, with just a little brown forming around the pit of some of the apricots, or it can be severe, in which case the entire crop can be lost. Compared with most other commercial fruits, apricots also have more years with a light crop. But the reward for a good crop can easily make them worthwhile.

Have you ever wondered what the apricot growers did with all the apricot pits? A lot of growers wondered what to do with them too until 1925, when Sewall Brown and Howard Scott opened the Scott and Brown Apricot Kernel Company in Los Gatos. It was located on Winchester Road near Vasona.[123] Scott and Brown dried the pits and cracked them to remove the kernels, which when dried have a flavor similar to that of an almond. They then ground up the kernels to make a paste and sold it to bakeries. The shells were converted into a high grade charcoal. In later years, probably in the early 1950s, apricot kernel oil began to be extracted for use in cold creams and other cosmetics. The oil was, and still is, highly prized, and demand for it has greatly increased the value of an apricot crop.

Pears and Peaches

Bartlett Pears and Cling Peaches were next, usually ripening in late July. The great majority of the pears and peaches were canned, but parts of both crops were dried in the same way the apricots were. Cling peaches are so named because the flesh of the fruit clings to the pit. Freestone varieties don't. Clings remain firmer when canned than Freestones and do not bruise as easily, so they are perfect for canning and accounted for the great majority of the Santa Clara Valley's peach production. In the Valley, both peaches and pears had more trouble with pests than most of the other fruits. Cling peaches, like apricots, are quite susceptible to brown rot. And, like pears and apples, the codling moths also have a fondness for them.[124] Leaf curl, caused by a fungus, is also a common plague of peach trees. It causes the leaves to curl and blister and greatly reduces photosynthesis. Sometimes leaf curl can be controlled by trimming, but that is very labor intensive and also reduces yield. The pear orchards probably suffered the most from infestations. Fireblight, a bacterial disease, was especially troublesome if there were warm spring rains. Near the end of the fruit growing era in the Santa Clara Valley, in 1957, an insect called pear psylla destroyed about half of the Bartlett pear trees in the state.[125]

Crop Pests and Diseases; Spraying

California has a long history of agricultural science, much of it born out of necessity. Along with all the varieties of plants that were imported into California came a large variety of plant diseases and pests. Many different crops started out well for some years, but were later devastated by harmful life forms that were introduced from other parts of the world. As early as the mid 1850s, desperate California farmers struggled to find ways to deal with crop diseases and pests. By the 1880s, California was a huge food producer, and the seriousness of crop pests and diseases caused great alarm. Many laws were passed, and from the early days inspection stations were set up at the ports of entry to try to control what was brought into the state. As crop disaster followed crop disaster, methods of dealing with the various life forms that destroyed crops became of paramount importance.

By the 1860s, microbiology had come of age, and the fame of Louis Pasteur, who had devised a way to kill the bacteria and fungi present in milk, spread around the world. Pasteur drew the conclusion that micro-organisms cause disease, and that was hailed as a huge step forward. Claude Bernard, another leading scientist of the day, who with Pasteur first developed the process later known as pasteurization, came to a very different conclusion. Bernard thought Pasteur went too far—that it was not germs that caused disease, but rather the terrain in which those germs existed that allowed the germs to thrive that was responsible. Bernard believed that the relatively stable state of equilibrium between the different but independent elements, or groups of elements, of an organism was what was vital.[126] In other words, he understood the importance of what is today called an ecosystem. Other top scientists also, who perhaps had more respect for the complexity and perfection of the creation, did not buy into the concept of eradication as a preferred means to treat disease. But Pasteur's methods were shown to be immediately effective, and from that time, the concept of "killing the pests" became the dominant "scientific" mindset. Scientists who disagreed were ridiculed as "non-scientific." Soon, in California and elsewhere, bushels of business and government money were being handed out to the agricultural schools to teach methods based upon the principle of eradication.[127]

To kill something, a toxic substance is needed, so lead, arsenic, and other arsenic compounds were used to control the codling moth; lime compounds were commonly used for fungus and insects, and kerosene and tobacco juice were used to kill just about everything.[128] Typically, a soap or oil solution was used as a vehicle for the toxic compounds. Oil was, and is, preferred because it will not wash off with water.[129] In 1883, a man named John Bean[130] purchased an orchard (probably in the Los Gatos area) that was infested with a kind of scale—an insect that sucks the sap of trees. It seems everyone else's orchard had it too. Treatments were known, but the methods of application—sponging and sprinkling—were inefficient and required far more of the toxic substances than was desired. Bean invented a high power, continuous spray pump to apply the pesticides, and the rest is history. The Bean Spray Pump, or others similar to it, soon became standard equipment on most of the Valley's farms.[131]

Prunes

Prunes, almost all of the French variety, ripened in the second half of August. With the exception of a bad infestation of a sap-sucking insect called thrips around 1910, prunes and most of the bugs got along pretty well.[132] But growing prunes did require some additional care not required with most other fruits. The branches of a French Prune tree are not nearly as strong as those of a cherry or apricot tree. In the early summer, when the fruit starts to come in, the branches of a prune tree need to be propped up so that the weight of the fruit does not break them or tear the tree apart. The props were wooden sticks of various lengths with a "V" shaped notch for the branch at one end. These were carefully placed under the heavily laden branches. In a "bumper crop" year, it was not unusual to see forty pounds of fruit on a small branch, and it was all one could do to keep the trees from falling apart under the weight of the fruit. Sometimes a tree would need a dozen or more props under its branches. If the crop was heavy, an orchard would be gone over and picked three or four times as the fruit ripened. If the crop was light, there may have been only one pass over the orchard.

(The first year I was on the Pitman farm, we had about ten pickers who in three or four weeks picked the orchard three times. The first two times the prunes were very thick on the ground, and the last time they were moderately so. The following year we had four pickers and were done harvesting in less than four days.) The variance in yield could have been for several different reasons, but it is not unusual to have a bumper crop followed by a very lean crop.

Prunes were not smoked as the apricots and pears were. There were two ways to dry them successfully without having them rot, grow mold, or attract flies during the drying process. The first way was to dip them in hot lye water for about a minute, then rinse them and set them out in the sun to dry. Dipping the prunes roughened the skin a little, which kept the fruit from fermenting at the pit. In the late

1880s, another method began to be used. That was to put the trays of washed prunes in a dehydrator. A dehydrator, also called a dryer or an evaporator, is simply an enclosed shed in which large fans are installed. The fans blow very hard and dry the prunes in about thirteen hours. When dried in the sun, the drying process took three or four days.

In the early days, most of the growers dried their prunes in the sun. However, there was some risk involved with this process. A rare, heavy rainstorm during the prune season of 1918—that filled the entire Santa Clara Valley with the pungent odor of fermenting fruit and required calling in military personnel to help clean it up—sent many growers into the waiting arms of the dehydrators.[133] Eventually, most of the Valley prune growers became members of the California Prune and Apricot Growers Association (which later became Sunsweet) and took their prunes to one of their drying facilities.

Walnuts

The walnut crop, the last orchard harvest of the year, ripened right after the prunes were in, usually beginning around the second week of September. Early pioneers in the Valley found an abundance of native Black Walnut trees near areas of natural drainage, just as they are now found throughout much of the northern part of the state. The Black Walnut has a most delightful flavor and is superior to other walnuts in some ways, but the work required to extract the small amount of meat from the thick, woody shell made commercial use of it not viable. English Walnuts were brought in, but these trees were susceptible to rotting from a fungus that grew at the base of the trunk. The problem was solved by splicing, or grafting, the English Walnut into the native walnut rootstock.

For a graft to be successful, the vascular cambium layers (the living tissue of the branches) must be aligned perfectly and kept alive while the union takes place. Both the cutting and the rootstock are cut at a long angle and then tied together. A dressing is then applied over the wound and kept moist to prevent the graft from drying out for the three or four weeks needed for the two to grow together.[134]

The native Black Walnut then became the rootstock for the English Walnut, and the industry thrived. This is why in California walnut orchards rough, black trunks extend about two feet above the ground, and above that the trunks are white and smooth.

The walnuts were shaken from the trees, picked up, and the green or blackened hulls were removed. The nuts were then set out in the sun to dry. Shuckin' walnut husks would leave one's fingers stained a deep purple color for a week afterward. After drying, the end of the nuts had to be checked for evidence of the codling moth. The presence of a small hole, a little sawdust, or a spider web-like substance at the stem end of a nut is sure evidence that the nut is infected. If a walnut grower is not willing to share some of his crop with the codling moths, he may need to spray—perhaps repeatedly.

After the crops are in, the farmer may or may not have a respite. When the weather turns cooler in the fall, the pruning must be done before the heavy rains come. Like the disking, propping, or the harvest itself, a farmer can't afford to be too early or too late. A farmer must pay close attention to the rhythm of the land.

The farmer's life is a most instructive dichotomy. The diligence, perseverance, and hard work required to produce good crops on a year by year basis cannot be exaggerated. At the same time, it is a life of total dependence. The rain must come at the right time and in the right quantity. The sun must be hot at the right time and not too hot at other times. The birds, insects, and other life forms are needed, but likewise must be kept in balance.

Although tremendous efforts have been made to lessen dependence: pesticides, imported water, new strains, and the most telling of all—crop insurance, in the end, the livelihood of the farmer stands out as a glaring example of the fact of human dependence.

It is no less so for all of us, but the example of the farmer may be the most dramatic. We need to exercise the utmost diligence, perseverance, and hard work if we are to produce anything of value, but while doing so, it is perhaps wise to continually, humbly acknowledge that there are many things that are beyond our control. There is One who gives and takes away, and upon whom all life depends.

Picking Apricots
Photo Courtesy, History San José

Prunes drying at Lester's Farm[135]
Photo Courtesy, History San José

Disking an orchard, circa 1920
Photo Courtesy, History San José

Trays of apricots going into a smoke house, or kiln.
Photo Courtesy, History San José

11
Olives, Grapes and Berries

Besides grains and orchard fruits, many other crops were tried in the Santa Clara Valley which met with varying degrees of success. It is one thing to be able to grow a crop in a garden; it is quite another thing to make a living off of it. Although nearly anything would grow in the area, the limitations of climate, soil, water, available labor, sustained good management, and transportation determined if there was a chance of commercial success. The story of the Arguello olive orchard is a case in point.

Olives

After the U.S. court's ruling on the Rancho Quito land claims, Ramón Arguello[136], the owner of what remained of the rancho, apparently began to receive payment from many people who had settled on what had been determined to be his land.[137] Having available cash, in 1865, he began to plant an olive orchard near the present intersection of Quito Road and Saratoga Avenue. Over several years, Arguello planted eighty acres of olives trees, put in a modern olive press, and built a packing facility.

Arguello's olive trees did very well, produced award winning oil, and received great publicity. These trees became the standard for the Mission Olive industry in California, and cuttings from them were sold to nurserymen and farmers all over the state and beyond. But as Arguello's olive trees matured, it became apparent that the trees were planted too close together. Olive trees can live hundreds of years and still bear, but they need adequate space in order to bear well. So Arguello pulled out every other tree. But the expense of pulling the trees, the now inefficient yield per acre, and new competition from other olive growers nearly broke the piggy bank. Supply had caught up with demand, and eventually almost all of the olive trees were torn out and replaced with walnut and apricot trees. One last hurrah for the famous Arguello olive ranch was that many of its trees were dug up and transplanted on the newly formed flat land of Treasure Island for the 1939 World's Fair[138]. The island is at the middle of the Bay Bridge between San Francisco and Oakland. If trees could talk, the few remaining olive trees on Quito Road would have quite a tale to tell.

Grapes

At one time, grapes were nearly as big a crop in the Valley as prunes. The grapes grown at Mission Santa Clara were not of a quality to produce commercial wine, but many settlers who came later brought their choice vines with them. As with fruit trees, nearly countless varieties of grapes were tried.

Elisha Stevens, who in 1844 led one of the first wagon trains into California, planted a vineyard in the foothills of what is now Cupertino. Stevens had come to California seeking for gold—before it was discovered. He was apparently trained in prospecting, and after leaving off his party, went into the Sierras looking for gold. He did not find what he was looking for, so he came back to the Santa Clara Valley and planted his vineyard. An interesting note: Stevens had been in many of the right places in the Sierras to find gold—even up the Feather River Canyon where some of the richest surface deposits were later found just lying in the river.

Stevens may have missed the gold, but he got the location right for his vineyard. The hill country of the Santa Clara Valley, especially on the west and south sides, was ideal for growing grapes. Many other settlers followed Stevens and planted vineyards in the area. Almost all of them were Europeans who had brought their own stock from their mother countries and were well-versed in wine-making. One of the earliest was Pierre Mirassou[139], who established the Mirassou Winery in what is today's Evergreen District of San Jose. Charles Lefranc, another experienced wine maker from France, established Almaden Vineyards in 1852 along the southern hills. After Lefranc's death, his partner and son-in-law, Paul Masson, went out on his own and started his winery in Saratoga. Masson's specialty was champagne, which his winery produced in Saratoga until the early 1970s. One of his original buildings still stands today; it is located off of Pierce Road and is now used for social gatherings.

There were countless others who planted vineyards. Even Ramón Arguello, of olive fame, did not put all his eggs in the olive basket. He planted grapes in the European style between his now widely-spaced olive trees.

Virtually all the grapes produced in the Santa Clara Valley were for wine making, and many varieties proved successful. Grapes for raisins were (and still are) grown in the Central Valley where it is hotter and more conducive to growing raisin grapes.

It is not so easy to just pick up a vine or a fruit tree, take it outside of its natural habitat, and plant it. Many things can, and do, go wrong. Just as many of the native peoples in the Americas died from European diseases because they lacked immunity to them, so it can be with plants. A case where something did go horribly wrong was when someone took cuttings from wild grapes in California and carried them off to Europe to experiment with them. A tiny insect, called Phylloxera, came along for the ride to which the European vineyards had no resistance. So the European wine industry took a big hit. And the California wine industry got off the ground.[140]

The California wine industry became huge, but the boom in the Santa Clara Valley was over by 1890. As more and more vineyards were planted and started bearing, the necessary infrastructure—cellars, presses, cooperage, bottle works, etc.—did not keep pace. The grape growers in the Valley were stuck with a lot of grapes that they could not process.[141] Vintners in other areas did put in the needed infrastructure, and because wine grapes cannot be transported very far, the San Joaquin Valley and the Napa/Sonoma area became California's wine country.

In the western and southern foothills of the Santa Clara Valley, many farmers tore out their vineyards and planted prunes and apricots. The Villa Montalvo property in Saratoga (now a cultural center) was purchased by Senator James Phelan in 1912 and had been almost entirely vineyards. Phelan tore out the vineyards and built his estate and gardens.

There are still many small vineyards in the western and southern hills of the Valley, and though they are really hobby-farms and not serious commercial enterprises, perhaps everyone who sees them appreciates that they are there. How pleasant is a vineyard! Other than these reminders of the past, all that remains of the grape growing era in the Valley are a few street names.

Berries

An account of the agricultural era of the Santa Clara Valley cannot be given without mentioning the strawberry industry. It is one of the major agricultural stories of the Valley. In the period up through the completion of the Trans-Continental Railroad in 1869, Chinese and American labor brokers, working for American mining and railroad interests, brought about 60,000 Chinese workers into California. They came from the Canton area in the Kwangtung Province in Southern China, and as such spoke Cantonese. Almost all were poor young men who were fleeing famine and domestic turmoil in Kwangtung.[142] Many, if not most, agreed to pay for their passage by working for one of the labor brokers after they arrived in California.[143] When the railroads were completed and the mines began to play out, many Chinese work gangs went to work in agriculture. They planted many of the early orchards in the Santa Clara Valley, and they also dominated the strawberry farming. Wealthy Chinese bought or leased property in the lowlands, mostly in the Milpitas and Alviso areas where artesian water was plentiful, and they hired Chinese crews to work as share croppers. Their formula was very simple: the owner supplied the land, plants and water, sold the crop, and took 50%; the workers planted, cultivated, and harvested the crop, and received the other 50%.[144] Some of the Chinese also worked independently for farm owners, while others leased land themselves and worked as tenant farmers. The newer arrivals usually worked in gangs supplied by labor brokers.[145]

As long as the Chinese were hired only for work that the Americans and newly arrived Europeans did not want—extremely dangerous work, stoop labor such as tending strawberries, or service work such as doing laundry—their presence was not only tolerated but even welcomed. However, with the completion of the major railroads, the labor supply outgrew the demand. The Americans would not work for the low wages or in the poorer conditions that the Chinese would, and as competition for work increased, many people treated the Chinese harshly. Others were more noble. In San Jose, members of the First Methodist Church established a school for Chinese children, and many other people had good relationships with the Chinese. But the church was destroyed by a mysterious fire in 1869, and a year later, San Jose's Chinatown burned down too. Despite a written death threat, the pastor of the church, Thomas S. Dunn, appealed to his congregation and the school was re-built within five months.[146] On the other hand, newspaper accounts stated that there was a lack of enthusiasm displayed by the firemen to fight the Chinatown fire.[147] Chinatown was rebuilt, this time with several brick buildings.

The economic depression of 1874 brought many more workers to California from the eastern states. When added to the recent immigration of Irish and the continuing Chinese immigration, the California labor market became saturated. Wages deteriorated, social unrest followed, and the Chinese bore the brunt of it. One result of the unrest was the Chinese Exclusion Act of 1882, which prohibited further Chinese immigration.

Within a short time, however, as the development of farms and orchards increased, there was again a labor shortage. So, in 1885, Congress opened immigration to the Japanese to fill the void.

One big difference between the Japanese and the early Chinese immigrations was that most of the Japanese came as families. The overwhelming majority of the Chinese had come as single men.[148]

The Japanese took over the vacated strawberry industry and were equally successful. Besides working the large operations, many Japanese workers earned enough to purchase their own small plots and grew strawberries on them throughout the Santa Clara Valley. Some of these plots were still being worked in the early 1960s. Many of the Japanese workers who came were very skilled in horticulture in their homeland and also performed much of the needed labor for the orchards. They also owned, and their descendants likely still own, many of the nurseries in the Valley.

12
Market Challenges and the Sunsweet Story

Bringing in a crop is only half the battle that the farmer faces. Securing a market for the crop that allows him to make a living for all his effort expended is the other half. Many things can go wrong, some of which have already been discussed in previous chapters. Fresh fruit must get to the market on time and in good condition or it is useless. In the early days of the fruit farms, many loads of fresh peaches or pears looked beautiful leaving the Valley but were downright disgusting by the time they got to Market Street in San Francisco. Dried fruit is much more forgiving, but still a lot can go wrong.

There is a vast difference between fresh or dried fruit on the farm and that fruit in the consumer's hand. A large amount of infrastructure, planning and labor go into transforming the former into the latter, and there are numerous pitfalls every step of the way.

When the railroads came in, that was very good for the farmers—a large part of the transportation problem was solved. When the railroads became a monopoly, another problem was created—the railroad owners could charge anything they wanted. They did, and American history is littered with the wreckage of farmers who became prey to a rail freight pricing structure based on nothing other than the customer's (or victim's) ability to pay.

Another pitfall is that a product can get a bad name, and a good market can be greatly damaged or even ruined. Quality control is a challenge in any industry, and the dried fruit industry is no exception. The quality of dried fruit can vary enormously. Sometimes small apricots that were dried too long or are covered with dust from the drying yard, making them almost inedible, are set out for the consumer. Likewise, a consumer will not forget the taste of a dried apricot that was picked too green before the sugars are developed. Nor will a consumer forget a box of prunes with dots of white mold on them. At times fruit like this gets dumped on a market. People buy it because it is cheap, then in the future decide to spend their money elsewhere.

Then there are the middlemen. Middlemen are usually necessary—someone needs to be the liaison between producer and consumer. But again, shortsighted individuals can greatly undermine the farmer's ability to make a living.

Of all the human-caused pitfalls a farmer may face, the market manipulators are perhaps the worst. These individuals—caring only for themselves—secretly gain control of a market and exploit everyone else involved. They can be absolutely ruthless and can do the economic equivalent of throwing a grenade in a crowded subway. Markets have been manipulated for millennia, but when market manipulation becomes parasitical and kills the host, that is quite another matter. The practice not only happened during the fruit growing era in the Santa Clara Valley, but also continues in agricultural markets to this day.[149]

Although new varieties of ways to cheat our fellow man are being developed daily, when looked at closely, they are all the same banana and are older than Jacob's Uncle Laban.[150] The Valley farmers experienced a lot of these "bananas" first hand, which led many of them (as well as farmers elsewhere) to the conclusion that the only true humans in the farmer-to-consumer equation were the farmer and the consumer. But the farmers were too small and those who controlled their products were too powerful.

The Grange Movement is an important part of American agricultural history. The Grange was formed beginning in the 1870s, when farmers all over the nation were banding together to try to protect themselves from low grain prices, high railroad charges, and political corruption. It took more than two decades, but the Grange did help to turn public opinion against the railroad monopolies and the political corruption of the day.[151] Likewise, several farm cooperatives and associations were started in the Santa Clara Valley, but most were short lived. The reasons for their failures were numerous, but probably the biggest factors were: inadequate business skills, inadequate representation in the industry, and last, the fox was guarding the hen house.[152]

The dried fruit industry is not as vulnerable as the fresh fruit industry since the product will not spoil as readily. But that will not cure the ills of a glut in the market. In 1890, the Valley produced

8,000 tons of dried prunes. Prices were high, and other farmers seeing this, planted more prune orchards. In 1900, 85,000 tons of dried prunes were produced, but the market had not expanded to accommodate the additional yield.[153] There had not been the necessary promotion and advertising. So what happened? Prices collapsed, and not a few prune growers lost their farms.

During the early years of the fruit growing era in the Valley, buyers from eastern cities came and made agreements with the local fruit farmers. The agreement could be anything that was acceptable to both parties. It could be made in February for $60.00 a ton for the dried fruit, or it could be made in August for $20.00 a ton, delivered to San Francisco. So, there was great variation in the deals different farmers got, and consequently a lot of "shenanigans" went on. Some farmers tried to time the market in order to secure the best price, only to see prices collapse later.

One thing in favor of the prune grower is that prunes are stored and shipped with an 18% moisture content, and they can last a very long time if stored properly. A prune grower can hold a crop for over a year if he wants to and perhaps get a better deal later. Or maybe not.

(By the way, your teeth will complain greatly if you try to eat a prune with a moisture content of 18%, but will be very happy with 25% or a little more. So before being set out for sale, the prunes are re-hydrated.)

To ward off the instability of market conditions, some larger fruit growers established their own brands and created their own markets. But most growers were not big enough to establish their own brands. And after all, they were farmers, not distributors. The only solution for most fruit growers was a working association, but as already mentioned, numerous farm associations in the Valley had failed. Most of the farmers had probably given up the thought that there really could be a cooperative organization that could protect them. Also, farmers tend to be very independent, which is great for getting farm work done, but disastrous when it comes to an orderly distribution of farm products.

By the time of Teddy Roosevelt's presidency (1901-1908), most people had had enough of the "Robber Barons" of American industry. Roosevelt is credited for breaking up most of the monopolies that had paralyzed the country, and for at least limiting the political corruption that ran many of the larger cities. These changes paved the way for Aaron Sapiro.

Sapiro became one of the leading agricultural attorneys in the world, and apparently he had a heart. He became so successful in organizing farm cooperatives that Henry Ford accused him of being part of a Jewish plot to corner the world's food supply. Sapiro sued. Ford apologized. Sapiro, instead of taking millions, would only take one dollar. And a lot of newspapers were sold covering the story.[154]

In organizing the co-ops, Sapiro insisted that they be formed along specific product lines. The Raisin Growers Association, the Citrus Growers Association, and the Prune and Apricot Growers Association, were some of the cooperatives formed under Sapiro. A trademark was needed that would be representative of California, so out came Sunmaid, Sunkist and Sunsweet. (How the Almond Growers Association ended up with Blue Diamond instead of Sunnuts, I don't know.)

The cooperatives, or associations, were organized along very simple lines. First, the growers would own the association and all of its assets. Second, they would elect a board of directors every year. Third, the associations would be as independent as possible. They would have their own dehydrators and packing plants, and set their own quality standards. Last, they would do their own advertising, establish their own markets, and split the profits, less expenses, among the growers.

The Prune and Apricot Growers Association (later Sunsweet) was formed in 1917, and within a few years most of the prune and apricot growers in the Santa Clara Valley were on board. It remained so until the end of the fruit growing era. At one time, there were thirty Sunsweet packing houses in the Valley.[155] Sunsweet was a well-respected name in the Santa Clara Valley for over fifty years. A previously unknown stability had come to many of the Valley's farmers.

Trays of prunes coming out of the washer; ready to enter the dryer

Next pages: packing prunes in one of the Sunsweet packing houses

13
Diary of a Cannery Worker

Lots of trucks stacked up today. The line's running out the gate and onto the street again. You would think that all the peaches get ripe on the same day! They'll probably ask me to work overtime.

Sun's not as hot today. A little dampness is still in the air. But then it's only ten minutes 'til seven.

For canning food you'd think it would smell better around here. But all that diesel and rubber could ruin anything. All that stale water down in the truck docks doesn't help either.

"Hi Marty!" Marty's one of the best men on the docks. He's got a heart. Not bitter. He just tries to do his job. It makes a lot of difference, you know, how the dock men are. They take the fruit boxes off the trucks and empty them onto the conveyor belt. They're supposed to empty the box slowly so that the fruit is spread out a little when it comes down the line. That makes it a whole lot easier on the sorters. I hope George isn't in a bad mood today! Jerk! He just dumps the box—so all the fruit is in a pile—and you have to scramble like crazy to keep up. Too much of that, and you have to complain to Harold. As far as foremen go, Harold's pretty good. But even he usually thinks you're whining if you say anything.

Glad I'm out of there! Got a job at a peach slicing machine. It pays ten cents an hour more, and you get away from George. Poor George! I feel sorry for him; he's not a very happy man.

I couldn't get off the sorting table last year—too many women on the list for a machine ahead of me. I wouldn't have got off it this season if Clare hadn't slacked off. They told her she still had a job, but couldn't work the slicer anymore because she was too slow. You've got to be ready with the next peach when the blade comes down, or it can get messy real quick—mangled peaches all over the place.

Timing's everything. You have to stand up straight and pay attention. Grab a peach, get it in place—you've got to line it up so it gets cut along the crease. Otherwise they look bad and don't fit in the cans right. And you've got to keep your fingers out of the way of the blade. Let the blade do the work. It's automatic—you just go with it. Don't move your hands more than you have to.

But it can bore you to death. I think the only thing that keeps you going is the fear of getting a finger sliced off. I suppose that's happened. They warn you.

The clanking of the machinery can lull you half asleep though. Got to stay alert.

I wish one of the mechanics would come and oil that bearing over above number four. It's been squeaking for a couple of days now. The drone of the motors you can get used to and don't mind after a while, but that squeak is pretty annoying. Boy, were the mechanics running yesterday when a belt slipped off over by the dipping kettles! After the peaches leave the slicing stations, they dip them in a solution there—lye water—to get the skins off them. From there they go to the canning room—over there. See that big tank way up there? That's the syrup. The peaches go in the can, the syrup gets squirted in, the lids go on, and then it's on to the cooker. It's all automatic now. Works good too—most of the time. Last Tuesday a hose broke, though, and there was syrup everywhere. Everyone else could stop when the whistle blew, but not the mechanics. I guess that's why they get paid so much. It took them a while to fix it, and they were cleaning up for hours. They had to hose down the whole place—no wonder the docks stink! It was sure backed up over there for a while. The forklift drivers were going nuts trying to keep stuff out of the way.

Oh that dear whistle! What a lovely song! It means we get ten minutes. Every two hours. Not ten and a half minutes, mind you. You'd better be back at your station when the machine starts up, or they'll get some one else who can be. We get half an hour for lunch. The first two hours aren't that long, but the last two…

I sure am glad I've got more to look forward to than working here the rest of my life! But it's not bad, really. Probably like everywhere else. I'm just here for the summers so I can get through college. I want to be a teacher. I like being with kids—especially the younger ones. They make me happy.

Most of the women here are older and like it well enough. They say it's worth it because they can draw unemployment for a month or two twice a year when the plant shuts down. After the summer fruit is done, there's nothing to do until the winter vegetables come in, and then there's a month or two before the cherries start. So they don't mind. They like getting paid for doing nothing for a third of the year.[156]

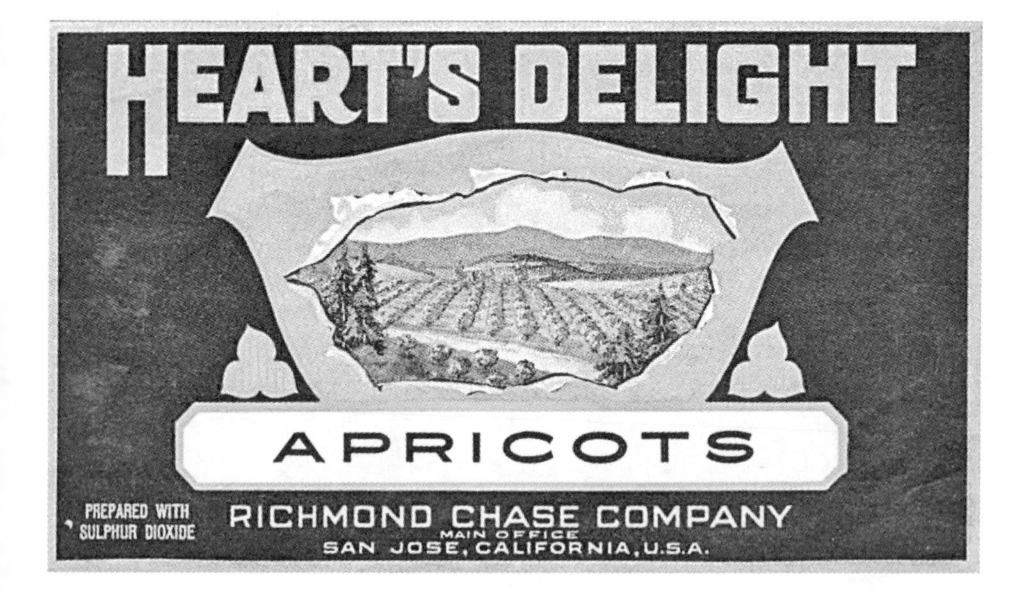

Cutting Pears
Photo Courtesy, History San José

14
The End of an Era

Chapter 9 gave a general history of the fruit growing era in the Santa Clara Valley up through the end of World War I. Chapters 10-13 filled in some necessary details, and this chapter will go back and pick up the rest of our history.

The growth and prosperity after World War I continued in the Santa Clara Valley and in the country as a whole through the 1920s. Then came the Great Depression. In the Valley, the 1930s were not as difficult as they were in many other places. Refugees, three hundred thousand strong, came to California from Oklahoma, Arkansas and the other surrounding states of "The Dust Bowl"—a Midwest plagued by drought. They traveled up and down the state in dilapidated old cars and trucks following the crops, and they set up tent camps wherever they could. Many came to the Santa Clara Valley and never left.

The Depression had been nearly global in its reach, and desperate nations began to do desperate things. Germany gave way to Hitler, Italy to Mussolini, and Japan to Araki and others of like-mind. President Roosevelt tried nearly every conceivable way to stimulate a broken US economy in the 1930s. The Golden Gate Bridge and the Oakland Bay Bridge were large and noticeable government projects in the near vicinity of the Santa Clara Valley; and as the war drums were beginning to beat, within the Valley itself, Moffet Field, with its enormous airship hanger, was built by the US Navy. The hanger was built for the USS Macon, a lighter-than-air aircraft carrier. But there were also government projects of a much smaller scale that were built in the area during that time. Some local roads were built, so were some of the small bridges over the creeks. But nothing solved the unemployment problem; that is, until the Second World War.

The War changed everything, as a wave of military bases and war material factories sprang up on a virtually unprotected West Coast. An area that had sent many of the "Oakies" and "Arkies" packing during the Great Depression, was now looking for all the workers it could get.

The War Productions Board enlisted several Santa Clara Valley industries to meet the needs of the war. Food Machinery Corporation, now FMC, in San Jose, was entrusted with the manufacture of landing craft vehicles. These were amphibious tank-like vehicles designed to transport troops and supplies from ships to land. Thousands upon thousands of troops were trained in them at Fort Ord in nearby Monterey Bay before going on to beach landings on the Pacific Islands. Likewise, Hendy Iron Works was put to work. Hendy, a leader in the manufacture of mining equipment since the 1850s, and a transplant to Sunnyvale from San Francisco after the 1906 earthquake, had seen very tough times during the Depression. The company had shrunk down to sixty employees and was in danger of being taken over by the bank. With the war, it blossomed again and employed as many as 11,500 men and women to build an array of war equipment including 754 reciprocating steam engines for the US fleet of *Liberty* supply ships. The canneries were once again going at full capacity, and there was no lack of work to do. The intensity of work kept up throughout the war years.

The coming of peace brought with it the huge challenge of unwinding war production and receiving back the ten million young men and women who had been sent overseas. This could not be done all at once. Not only were there insufficient ships to transport everyone all at once, and insufficient docking spaces for all the ships, but the orderly processing of the soldiers and sailors itself took a great deal of time. Perhaps the biggest problem of all was that in most of the cities where these servicemen came from, there was a shortage of jobs and housing. During the war, other people, mostly from rural areas, had moved into the various manufacturing cities to do the war materiel work, and housing was now scarce. So the entire process of receiving back the servicemen was slowed down in order to try to iron out the logistics.

Many servicemen, or "GIs"[157] as they were called, had to stay put far from home for an extended period of time. Even those who made it back to the West Coast could not just take off for home. They had to "get their papers," and that could take months. In the true military spirit of, "Hurry-up and wait," which these men were well accustomed to, they would be required to check in daily for what usually turned out to be no reason at all, other than for the purpose of keeping them nearby. Although the great majority of these detainees could not go home for a visit, they were on occasion given short "leaves."

OK, you're away from home, you can't go home, and you're a little tired of being around things military. What better to do than to go have a look around? For many GIs stranded on the West Coast, the more they looked, the more they liked what they saw. Many never went back east at all. Others went back just long enough to grab a wife and tell her, "Honey, start packing!" Still others went home to Detroit, or Pittsburg, or wherever, shoveled snow for the first time in a few years, were frustrated by the lack of employment they found, or finished an education provided through the "GI bill," and then said, "Honey, start packing!" They had seen a Good Land and they headed out to make a go of it there. The history of California a hundred years previous repeated itself on a much larger scale.

Of the thousands of former GIs who came to the Santa Clara Valley after the war, a large percentage had been stationed, at least briefly, at Moffet Field, Fort Ord, or other nearby bases. While stationed there, many had laid plans for how they could make a go of it in the Valley after the war. It was of little consequence that the place was nearly all orchards. The West Coast was ready to be developed and there was sufficient work waiting to be done.

In the Santa Clara Valley and in some other places on the West Coast, the post-war migration of GIs took on a fever pitch. With low-interest GI loans allowing servicemen to purchase a home with nearly nothing down, a building boom took place. These government loans were not based on wishful thinking or fanciful idealism. They were based on the fact that most of the rest of the industrialized world had been bombed up, and there was good reason to believe that US manufacturing would be very dominant for at least several years while the war-torn countries were recovering.

The housing tracts that grew up in the Santa Clara Valley during that time were of the assembly line type, and many were in places far from San Jose, such as in Saratoga or Cupertino.

As late as about 1956, Saratoga Avenue was still a two lane road where it intersected another two lane country road named Stevens Creek Road. The only structure on that corner was a bar and another small business. All else there, and nearly all the way into San Jose from Saratoga, was orchards. Nearly everyone who lived out that way had to go into San Jose for work, and for many the best route was to drive down Saratoga Avenue. One man saw opportunity there. He bought a little piece of ground from an orchard owner on Saratoga Avenue and put in a hamburger stand. Another enterprising individual took a larger bite out of another orchard just up the street at Payne Road, purchased a large army surplus Quonset hut, and opened a "Supermarket." The pink and purple Kaiser gravel trucks that came roaring down from the quarries in the hills had written on their doors and back panels, "Find a Need and Fill It," and many enterprising individuals were doing just that. Kaiser identified and articulated the driving force behind this new development of the Valley.

The Santa Clara Valley was not encumbered with many of the problems that plagued older cities back east. It was virgin land, close to natural resources, had adequate infrastructure in place, and now was well know as a viable place to live. At first, mostly smaller businesses moved in. But then Ford built an automobile assembly plant in Milpitas in 1955, Lockheed moved its Missiles and Space Division to Sunnyvale in 1956, and IBM built its San Jose Research and Development Laboratories near New Almaden, in 1957. The industrial rush was on.

There was one good thing, however small and fleeting, that came with the loss of the orchards: free fruit. Word got around where orchard after orchard was being abandoned, and people went out picking. It was not unusual to see an abandoned apricot orchard not only full of weeds, but also full of cars and families on a Saturday afternoon in July.

The development that took place in the Valley was absolutely unbridled. San Jose's city manager and other city leaders wanted to make San Jose, "The Los Angeles of the North," and to a certain extent they succeeded. At that time, state law allowed municipalities to annex land with fewer than five residents per acre without voter approval. All through the 1950s, the City of San Jose's City Manager's staff was very busy "persuading" orchard owners to allow their property to be annexed, and redrawing the map of the city. San Jose did not draft a General Plan until the 1960s.[158]

The rapid annexation of land by the City of San Jose alarmed many people in the surrounding communities. To keep from being swallowed up by this machine, the people of Campbell (in 1952), of Cupertino (in 1955), and of Saratoga (in 1956), voted for independent incorporation. But due to ever increasing property taxes on the farm land, even the smaller cities could do almost nothing to stop, or even slow, the development trend.[159] The flat land, which is easy to build on, was built up first. It was not until many years later, when the suburban sprawl had entered the foothills and the Valley was commonly filled with smog, that the political climate finally changed. No doubt, had the leaders of San Jose had their way, today the hills of Los Gatos, Saratoga, and Cupertino would have all the beauty of those in Daly City or South San Francisco. It seems that some civic leaders had the mindset that there was nothing more beautiful than a new tract of homes, a new strip center for shopping, and more paving.

All this was in a valley that John Muir had once described as, "One of the most fertile of all the coastal valleys". Within twenty-five years of the war's end, nearly the entire Valley was covered. Except for a few small orchards here and there—none self-sustaining—the fruit industry that had reigned over the Valley for the previous seventy-five years was, for all practical purposes, gone. The "Garden of the World" was laid to rest under a blanket of asphalt and concrete.

Photo: An old prune tree. The trunk has been largely eaten away by termites, but the tree is still bearing well. This was not unusual.

Part Two

The Story of One of the Last Farms

15
Early Days of the Farm

During the economic recession of 1874-1875, many easterners moved to the Santa Clara Valley. A good number of them were farmers from the Midwest, and among them was a Tennessean named William Rice. Shortly after arriving, Mr. Rice purchased 190 acres of land on the western edge of what had been Rancho Quito. The dry, brush-covered land that Rice purchased was on the west side of Saratoga-Sunnyvale Road in the low hills a mile north of the town of Saratoga. At the time Rice bought the property it was considered almost worthless. By the time Rice arrived, there was more than a twenty year history of many failed ventures in less desirable parts of the Valley, and many people scoffed at yet another foolish newcomer wasting good money and a lot of hard work on another hopeless prospect. Rice, however, saw it otherwise. He may have been a newcomer to the Valley, but he was not a newcomer to farming and farm development. In his native Tennessee, he had steadily built up and developed about 600 acres of farm land. He came along at the time when the railroads had recently been completed, and put a crew of unemployed Chinese workers to work clearing the land. As the Native Americans had provided the labor for the missions and the early ranchos, at that later time the Chinese made up a significant part of the labor force in the Valley.[160] After clearing the land, Rice's workers then planted prune and pear orchards. The trees rooted quickly and in about seven years were producing very well. Once these orchards were established, Rice, now in his sixties, began selling off portions of the land, mostly to other incoming Midwestern farmers. By the late 1880s, he had sold off all but about thirty acres.[161]

In 1886, Rice sold twenty-six acres of his developed land to another Midwestern transplant named Lyman McGuire. The next year McGuire purchased another parcel of land and a Fleming Fruit Dryer, and opened a packing house where he handled not only his own fruit, but that of many other local farmers as well. In 1887, he packed and shipped 110 tons of dried prunes, apricots and peaches in addition to his own fruit. The dried fruit industry was booming at that time, and the packing house did so well that he was able to have a new home built in 1893. There he and his wife, Ella, raised three girls.[162]

In 1900, Mr. McGuire was the secretary for the first Blossom Festival in Saratoga.[163] The Blossom Festival came into being after the drought years of 1897 and 1898, during which time the soil became hard as stone, the fruit trees dried out and cast their upripened fruit to the ground, and dust covered everything. The humbling of a drought to a farmer cannot be exaggerated—he can do everything right, but if no rain is sent, it is all for nothing. In March of 1899, the drought ended with an abundant rainfall that lasted three days. The relief to the farmers can only be imagined, and a special meeting was called at the Saratoga Christian Church for the purpose of expressing thanksgiving. Everyone was invited and the meeting was well attended, even though it was still raining and there were no paved roads at the time. Many farmers expressed their sincere gratitude if in no other way than just by the fact that they came. After the meeting, an old retired minister known as "Sunshine" Williams, full of thanksgiving, thought it would be fitting to have a celebration and invite city folk to see, "The God-given glory of springtime" from the Saratoga hills. His idea was to have an old fashioned picnic in a place where there would be a commanding view of the Valley full of blossoming fruit trees.

Writing about this gives me chills—if you have ever stood on a hilltop looking out on such a scene with a gentle breeze caressing you with the fragrance of a million blossoms, you will not likely forget it. Many city people came and were in awe. Most had never imagined such a sight.

The next year's festival was larger, and so on through the years. In time, the reason for the festival was perhaps overshadowed by grandiose pomp and ceremony, but it was a Saratoga event for more than forty years.

After some years, Mr. McGuire also planted an apricot orchard and more prunes in the back acreage of the farm. In later years, he was also one of the first to sponsor the Sunsweet Association.[164]

When the McGuire's flower faded, along came Robert and Mildred Pitman.

Robert was born in 1901, the middle child of five children and the youngest boy born to Homer and Anna Pitman. At that time, the Pitmans were in Ohio where Homer attended seminary. After his studies were completed, Homer took a pastorate in a Presbyterian church in Modesto, California where the family lived on a small farm. When World War I came, Homer was called to serve as a chaplin and was sent to France. Like so many other families at that time, the Pitmans were in transition. The oldest son, John, also went off to war, and the rest of the family went to live with Homer's sister, Mary, in Los Angeles. After the war was over, Homer accepted a pastorate in San Francisco, and the Pitmans re-united and moved north.

After the Pitman family settled in San Francisco, son Robert became disinterested in his schooling. What he wanted was to work with his hands, so he went to work for James H. Pinkerton Plumbing Company on Howard Street. Mr. Pinkerton knew Homer Pitman and saw potential in Homer's son, so he put him in the office. This was not exactly what the young man wanted, but it must have turned out alright because he ended up marrying the boss's daughter, Mildred, who also worked in the office.

In course of time, the Pitmans had three children, Bob, Mary, and David. In 1938, when the youngest, David, was five years old, the Pitmans decided to go farming. In Modesto, farm living had gotten into young Robert's blood, and the move was a natural transition to make at the earliest opportunity. For him. Mildred, however, was a San Francisco lady—of the hat and gloves variety. But that wore off. In time.

It did not matter that Robert Pitman knew little about farming. He would learn. And so he did. Shortly after arriving on the farm he bought a milk cow. The problem was, he didn't know how to milk a cow. So he went to his neighbor, Mr. Panetta the furrier, and asked him if he knew how. Well, no, as a matter of fact he didn't, but he said they could learn together. And they did. Mr. Pitman took one side of the cow and Mr. Panetta took the other, and they began to milk the cow. But cows are very particular about the way they are milked, and they want it done in the same way every time. No exceptions. So when Mr. Panetta took up his station on what was considered by the cow to be the wrong side, he got a swift kick that sent him flying. They both learned something, though the lesson was a little tougher on Mr. Panetta.

When the Pitmans purchased the farm, it was leased out by the McGuires with two years remaining on the lease. Leasing of farms, or portions of them, is a common practice nearly everywhere that there are farms, and the Valley was no exception. People who had worked the land for their whole life had a hard time selling out when they could no longer do the work, so they often leased the land to someone who could work it. There were plenty of farmers who leased these parcels, and the arrangement usually worked well for both parties. Many farmers farmed several detached parcels wherever they could lease them.

Those last two years that the farm was leased out gave Robert Pitman time to observe how fruit farming was done. A kind farmer across the street, Steve Williams, also taught him many things, and Mr. Pitman learned how to farm. The Williams had no children of their own and took these younger greenhorns from the city under their wings.

The house that McGuire built had stood empty for some years and was in a state of disrepair, but that was no problem to hardworking Robert Pitman. The house had been built on posts and piers and he replaced them with a new brick foundation. He added a bathroom upstairs, removed the interior pasteboard wall covering and replaced it with plaster, and put in hardwood floors throughout. In a word, he made the house like new.

Chickens and a small flock of sheep were added, and with the milk cow, the animal husbandry part of the farm was now complete. The main cash crops of the farm however, were apricots and prunes. Many of the farms in the area had both apricots and prunes. The fruits ripened at different times and by having both, a farmer not only hedged his bet against a bad year for one crop, but also found it easier to attract workers because he could offer them work for most of the summer.

The Pitman farm was dry-farmed, that is to say, the orchards were not irrigated. Irrigation was not necessary because the nutrient-rich soil was a sandy loam and as long as the orchard was disked properly every spring, the rainwater would soak deep and the tree roots would follow. Besides being less work, dry-farming had the great advantage that there was no need to spray for insects which came out in the spring and were attracted to irrigation water. The "worthless" land that William Rice had chosen was as good as it gets for growing prunes.

After being on the property for a short time, Mr. Pitman became interested in the parcel of land in back of his farm. It had a prune orchard on part of it, but no house. He asked around to find out who owned it, but it seemed that no one knew. After a diligent search, he found that it was in the hands of Uncle Sam who was short of cash. Mr. Pitman sent him a check for $1,800.00 for it, the check was cashed, and the papers followed in time. Not a bad investment for 19 acres in the Saratoga hills. The parcel, currently the Surrey Lane properties off of Pierce Road, was then added to the Pitman farm and was called the "L" because it took a turn near the back of the property. Mr. Pitman left the oaks that were on a knoll, and on the remaining unused land planted "The Little Orchard," as it became known, in prune trees.

So the Pitman family settled in to farming life, and being kind people, they became good neighbors. Except for the cow. Cows cannot be trained in certain respects, and one morning when the electric fence was not functioning properly, and the grass on the other side was in fact greener, Nellie decided it was indeed a very good day. The neighbor did not think so, as she awoke to see a happy cow polishing off the last of what had been her vegetable garden. This resulted in a sheepish Mr. Pitman calling his insurance agent and telling him something to the effect of, "Umm, I have a problem…"

The Pitman home. Built for Lyman McGuire in 1893

16
Growing Up on the Farm[165]

Work

On a farm, as everywhere else, things are always breaking down, weeds are always threatening to take over, and food does not just jump up out of the ground and into your mouth. The difference is that on a farm when something breaks down, you have to fix it yourself; when the weeds grow up, you have to take care of them yourself; and if you want to eat, you have to grow the food yourself.

Farming is work. And lots of it. On the Pitman farm, it didn't matter who you were, there was plenty of work for you.

Dave Pitman, the younger son, told me, "Even the cat worked. Topsy, the cat, was not fed, other than an occasional bowl of milk. Her job was to catch gophers and mice, and she was good at it. She had to be. On one occasion, she disappeared for about three weeks. But she came back. She must have decided that it was alright there after all. I suppose she wanted a better life for her children though, because after her kittens were a few weeks old, she would go deliver them to some of the neighbors.

"When I was about six, during the apricot seasons, it was my job to pick up the 'grounders.' If an apricot fell off the tree and was bruised, and half of it was still good, the good half was used. The rest went to the cow or to the sheep. Nothing was wasted."

Mary, the daughter, helped Mrs. Pitman in the house. She was a great prune picker too—she often picked more than 20 boxes in a day, which was no small feat.

The oldest child, Bob, grew into the job of farm mechanic. His dad saw that he was mechanically inclined, so he bought him a complete set of tools and encouraged him. Bob also did the disking.

I was told that if you ever visited the Pitman farm, Mr. Pitman would put you to work. Maybe that's why the migrating flocks of geese and ducks that passed overhead every winter never stopped—even when the pasture was flooded. They were probably afraid they'd be put to work too.

Everything was done the best it could be. There was no other way. It took four people to plant a tree. "Dad would dig and plant, and three others would line it up by sighting down the rows of trees," said Dave. Likewise, all were responsible. Dave told me the story of how he was goofing off one time and was pushing a cart full of apricot trays too fast. The cart couldn't make the turn from the shed to the kiln and went over sideways spilling the whole load of cut apricots. The trays measure three feet by eight feet, and at age seventy-six he still remembered it was *seventeen* trays full of apricots he had to pick up. His dad, though he saw it all, never said a word. He didn't have to.

War Time

During the war most goods were rationed. There were ration cards for shoes, meat, sugar, and many other goods. So people traded ration cards with their neighbors to get what they needed. Mrs. Pitman traded sugar cards for shoe cards. Farming is hard on shoes, and the kids needed shoes. One big advantage for the farmers was that there was no gasoline rationing for them. Farmers were considered part of the war effort—they needed to bring in the food. Everyone else was rationed, sometimes severely.

It was during the war years that Mildred Pitman was transformed from a city lady into a farm wife. Both of her boys told me, "She could squeeze a nickel until the buffalo hollered."[166] Mrs. Pitman did not do much canning except for string beans, a Pitman staple. Sometimes neighbors, happy to receive fresh food, brought over food they had canned in return. By working together in this way, everyone made it through the war years.

Dave told me, "We didn't use veterinarians. It was 'heal thyself.' If the dog was sick, he would be allowed to stay in the kitchen at night and have some milk, but that was as far as it went."

After Mr. Pitman had been farming a few years, his father told him, "It's time to stop this farming stuff and go back to work." His logic was that it was easier to make a living in San Francisco—one did not have to work so hard.

Too late. Robert Pitman was a farmer.

School

For the Pitman children, growing up was not just a matter of farm work. A formal education was needed to balance the valuable life-education they were receiving, and their parents were fully committed to providing it. Dave told me, "After dinner, Dad and Mom would wash the dishes while we did our homework at the kitchen table. Dad washed, Mom dried. While putting the dishes away, sometimes she would tell him, 'Little pitchers have big handles.' That was a code that we cracked. It meant, 'Be careful what you say—the kids are listening.'

"Through eighth grade, school was uptown on Oak Street about a mile and a half away. The school bus came by the many farms to pick up the kids. We were first on, last off. After picking us up, the bus turned left on Pierce Road, then went up Mt. Eden Road, and came back down Big Basin Way. On such a drive through the hills, constantly shifting gears and without the aid of power steering, the bus driver was not about to put up with any misbehaving. The first offence would see to it that you were put under the back seat. If there was a second offence, you'd be let out. And it could be a *long* way to walk. There was rarely any trouble."

The high school was in Los Gatos, about five miles away. The only bus to the high school was the Peerless Stage Bus which the kids caught at the depot in Saratoga. In those days, the road between Los Gatos and Saratoga was a winding country road, not the straight highway it is today.

Home Life

Home life was simple. Dave told me, "Mom played hymns on the piano. That was what she enjoyed. All the kids took piano lessons, and it stuck with Bob.

"Before television, the farmers would go visit each other and just talk. I learned a lot during those times."

Though the farmer's work was never done, there were priorities. Mr. Pitman read to his children. He read mostly stories from the Old Testament of the Bible. He talked to them about the people in the Bible and about what they could learn from their experiences. In later years, when he taught an adult Sunday School class, this was the first thing he emphasized to younger parents—"Read to your kids—read them Bible stories."

I asked both Pitman boys, Bob and Dave, "What did your Dad emphasize most in bringing up his children?" They both had the same answer, "Integrity." Bob added, "Every human being needs to be treated with kindness and encouraged. You don't walk on people; you pick people up. Have no favorites. Treat everyone alike and with respect. Above all, be honest."

To Mrs. Pitman also, honesty was all-important. Dave told me that when something happened among the children that shouldn't have, and no one would admit that they had done it, his mother would respond by saying, "Alright, this happened, but nobody did it. I declare this a mystery and I will never bring it up again!" And she didn't. What she did do, was tell the children how happy she was that she had perfectly honest children. And she did not waste an opportunity to tell neighbors and friends so—in front of the children. Dave said, "After about three weeks of her doing this, we just couldn't take it. It didn't matter if you did it or not, you wanted to confess it just to get her to stop!"

The worst memories of growing up on farm? Bob told me, "It was life—you did what needed to be done.

"One of the best things was the support I received from my parents—the modeling of life. I was impressed by how my father handled his problems. He was not quick to make decisions and he had a level of common sense. I also appreciated the help with my homework."

I asked Bob, "What was the most valuable thing that you learned from your dad?" He replied, "The value of work. There is a role for each of us in life to which we should be committed. We should be creative and energetic to carry it out."

Next pages: The back acreage of the Pitman farm, about 1959. Note the new houses on the hills on either side of the farm. A plot plan, or map, of the farm is in the Appendix.

17
Cock Sure and Turkey Milk

Growing up on a farm is a little different. There are a lot of characters, and not all of them are people.

When the Pitman children were growing up on the farm, there was a large old barn down the hill in the back of the property. It was home to the chickens, and when it rained the cow was brought in. The self-proclaimed owner of the place, however, was the rooster. He would attack all trespassers. The dog, the kids, anybody. Dave told me, "He attacked my dad one day. That was a bad move. A well placed kick made a change in his mind. It seems he came to the conclusion that, 'I can share the barn with this guy.' The rooster died when he decided to take on a coyote who got in."

The Pitmans always had a milk cow when the children were growing up. Usually there was just one, but sometimes there was a nursing calf also. Calf birthing is not always automatic—sometimes things go wrong. Once a calf being born got turned the wrong way and Mr. Pitman had to go in and turn it. He was alone down at the barn when he saw the cow in trouble, and there was no time to go up to the house to get help. He reached in and got the nose out, and the calf started breathing, but the birth would not progress. For nearly half an hour he hung on to keep the calf from getting sucked back in and suffocating. He came back up to the house totally exhausted and wet, but happily relieved. He had saved the calf.

For a short while there were two milk cows. But there was just too much milk and Mr. Pitman, not being one to waste anything, bought a bunch of turkey chicks and raised them on the extra milk. I'm told that some of the hens grew to thirty pounds, the toms to forty. Word got out, and people were interested in the turkeys. (This was in the days before cholesterol became a household word.) One man came out to see the turkeys, and couldn't believe their size. He grabbed one, held it up by its feet, and milk came out of its beak. He was delighted and bought them all.

He said the meat was buttery. I guess so! Maybe that's where the Butterball Brand of turkeys comes from, I don't know.

18
A Drunk Cow and a Kleptomaniac

All farm animals are individuals and have thought processes and personality traits like we do. This chapter is comprised of a couple of cases in point which were related to me by Dave Pitman.

Milk cows need attention. They like to be talked to, petted and loved just like the rest of us do. They can be kind and affectionate; and they can also have an ornery side, or a glitch in their behavior that becomes trying.

Such was the case with Nellie, the Pitman's first milk cow. Mr. Pitman, or the oldest son, Bob, would usually milk her out in the pasture in a mutually agreeable spot, but on rainy days the milking needed to be done in the barn. The dirt floor of a barn with cows in it can get downright sloppy. To control the mess, the farmer throws straw on the floor, which serves as a sponge, and the straw is mucked out and replaced as needed. But if there is a period of continual rain and the cow is barn-bound, it can get ugly. And so it was one rainy morning at milking time, and both Mr. Pitman and Nellie were uncomfortable and not particularly happy. He sat down on the milk stool to milk her and Nellie gave him a little nudge with her hindquarters that sent him sprawling into the muck. Not pleasant. Mr. Pitman went up to the house, changed clothes, and came back to resume the milking. But Nellie must have found the previous scene amusing, so she pushed him again and obtained the same result. Unlike Nellie, the farmer was not amused. So this time after changing clothes, he stopped at an apricot tree on the way down to the barn and cut off a switch. Instruction was given, the lesson was learned, the milking was finished, and the behavior did not continue.

There was another time, when for some reason, Nellie had some kind of an attitude problem, so she was chained up to an oak tree in the back acreage with a bucket of water. The kids were not allowed to go visit her. She had food and water, but being exiled was more than the social Nellie could bear. Her pleas for mercy were not heeded for a few days and the kids wept for her too. The neighbors also got their fill of a crying cow, but after she was received back she was as nice as could be. Tough love.

At the Pitman home, Sunday was not a day of work. If the apricots were out and would get too dried by being out all day, or if it looked like it would rain, they would stack the trays, but that was about it. Sunday was a day at church and in the home.

Unusable apricots, or the apricot slop as it was called, were always fed to the cow. One Saturday before the 4th of July, for some reason this was not done. The next day was Sunday and there was no work, and the following day was the holiday, so they did not work then either. So the cow didn't get the apricot slop until Tuesday. The apricots had fermented by that time, and the next morning Nellie was so drunk she was staggering. Her tongue was hanging out to the side, she was drooling profusely, and her eyes were bloodshot. She was downright loopy. It was hard to milk her too, as she had a hard time keeping her balance. The milk was thrown out—Mr. Pitman didn't want anything to do with it.

Poochie was the Pitman's first dog on the farm. He was as friendly as could be and you couldn't help but love him. Chips, the next dog, was different. He had his own agenda, and it didn't involve you. The kids were grown and gone when Chips came on the scene, and because the kids were gone, there was not the need for as many chickens. But there were still a few chickens left and Chips killed one of them. He knew he was in trouble, so he buried it. But I'm told he didn't do a very good job.

Chips had a habit of picking on the animals—he'd nip at the sheep's heels or otherwise harass them. One day, he got cornered by the ram, and all the ewes quickly joined in. It was pay-back time. They would have killed him had Mr. Pitman not gone in and gotten him.

When the Pitmans went on vacation, they would leave a bucket of dog food and another bucket of water, with the faucet dripping into it, for Chips. Then off they would go for a week or two. During these times Chips turned into a kleptomaniac. He would go and steal stuff from the neighbors—a dog's dish, a small rug, clothes off the clothes line, a small garden tool, anything. When the Pitmans got home, there, next to the tank house would be a pile of his plunder. It was embarrassing, but what are you going to do? They would put Chips' loot on the fence out front and call the neighbors to tell them if they were missing anything to please come by.

19
A Moo at the Door and the Mayor's Dog

The kids grew up and went off on their own, and certainly the Pitmans, like all empty nesters, had some of that inevitable questioning if they missed anything—if somehow they skipped an important lesson. I'm sure that at times they wondered if the methods they used to train their children worked and if the children understood. But at least worrying about the kids coming in too late was diminished.

Dave said, "Dad never cared what time we came home, because he was the one who picked the chores for the next day. If I was out too late on Friday night, there would be a ditch waiting for me on Saturday morning. He would never say a word about my coming in late; it was just, 'Oh, we have some digging to do.' I wasn't stupid."

Dave told me about when he would come home from college for a visit. "Before I left, Dad would get a chicken, chop its head off, put it in a bag, and hand it to me. I'd have to take the feathers off and do the rest myself!

"About the time that I left for college, Dad got his last cow. That was Sally, and she was really just his pet. She would come to the back door to be milked every morning at eight o'clock, and she followed him around like a dog. But they really didn't need the milk anymore, so eventually he traded her to Mr. Peake, the dairyman, for a beef animal. Dad said the meat was tough. But then Sally was not accustomed to going to a stall to be milked like all the other cows, so Mr. Peake had to go out and get her every morning. I guess it was an even deal."

I asked Dave how Sally was kept out of the garden and out of the orchard. The answer was the electric fence. Electric fences are very effective for keeping livestock and other animals out of an area. The saying, "Once bit, twice shy" really applies. It is a rare animal that needs to get zapped twice before it knows to stay away. The electric fence on the Pitman farm was just two wires, one was about a foot off the ground, and the other was about two feet off the ground. The wires were about an eighth of an inch in diameter and were strung between short steel fence posts that were very widely spaced. The voltage was very high, but the amperage was extremely low, so it packed a wallop but did not injure. The electric fence could be moved or added to easily, and in addition to protecting the garden and the orchards, it was used to divide the pasture into quarters so that there was always fresh grass for the sheep. It was also used on the perimeter of the farm in the back acreage.

If the grass grew too high and was not cut back, it would ground out the fence; but otherwise the fence worked very well almost all of the time. If the fence was turned on or off was of little consequence to an animal that had once touched it when it was on. That animal knew to keep its distance!

As the population of the Valley grew, more and more houses were built around the farm. There were dogs at many of these new houses, and at least one of the new neighbors didn't want to build an adequate fence to keep Fido in, but did object to him being bitten by that evil wire. The problem was that the dogs could see into the farm and would watch the sheep. And dogs are dogs; and sheep are sheep. It was a classic case of where farming meets development. And where that happens…well, from what I've seen, the farm typically doesn't last too long.

Now, it is a rare dog owner who will admit that his or her dog is a problem; but the fact is that sometimes they are. The pack instinct is also well preserved in domesticated dogs. If given the opportunity, it is not unusual for dogs to travel around in a pack at night. Typically, the dog owners are the last to know it. It was hard for some of the new neighbors to understand that their nice German Shepherd or Labrador Retriever that played with their toddlers during the day, roamed the hills at night looking for trouble.

All this being the case, sometimes a pack of dogs would come onto the farm, and once a pack somehow got into the pasture and killed a lamb. Mr. Pitman heard the commotion, went and got his rifle, and shot the biggest of the bunch.

It ended up being the mayor's dog.

20
‘Cots

I chuckle every time I hear someone say, “Lōs Gatōs.” To refer to the town in that way is to make a declaration that you are a newcomer and not accustomed to the local dialect. Everyone I knew when growing up referred to the town as “Lasgadus.” This gringoized pronunciation is disappearing in the area, but I’m afraid many of us will go to the grave with it. Likewise, we did not usually refer to apricots as apricots. They were simply ‘cots.

In the early 1960s, the labor laws were such that at twelve years old one was allowed to do farm work in California. However, a work permit had to be obtained first. So, like many other kids, I went down to the junior high school that I was to attend the following year, filled out the paper, and got a ticket in my hand to a realm of new adventures. Getting a work permit was similar to obtaining a driver’s license. With it, certain opportunities were opened up. It was no matter to me that the card stated that my eligibility was at a future date, the date when I would turn twelve. As soon as I got it, I determined to start looking for work so that I would be ready when that time came. Leaving the school and peddling my bike in the direction of home, I went looking.

The first place I stopped was at a large wooden shed at the edge of an apricot orchard where there was a fair amount of activity going on. Some kids were outside unloading boxes of apricots from a truck, so I walked over to them and asked who I should speak to about going to work. I was pointed to a young man who asked me, “Do you have a work permit?”
“Yes”
“Come over here.” He led me through the shed which, coming in out of the bright sunlight, seemed quite dark. The shed was full of kids, all older than I, who were standing at what appeared to be tables cutting apricots and placing them onto wooden trays. The young man led me over to an unoccupied space at the back of the shed, put a large wooden tray over four wooden fruit boxes that had been stood up on end, and had someone else bring a box of apricots. The box arrived and was emptied onto the tray. The young man then cut a couple of apricots to show me how it was done. He cut

the apricot along the seam, dropped the pit into a coffee can, and then placed the halves cut-side up on the tray, starting in the corner. He then told me it was fifty cents a tray, handed me the knife he had used, and walked away.

I remember thinking something like, "Uh… wait sir… you see I'm really not eligible to be here yet. I was just wanting to…" But nothing came out of my mouth, he was gone, and I was left staring at a tray of uncut apricots. Other kids were cutting theirs and filling up their trays, so I watched them for a moment or two, then picked up my knife, and began my career as a 'cot cutter.

In those days, there was not a five part form that had to be filled out in order to begin the employment application process so that one could, after following through with the various steps of the procedure and filling out all the other needed forms, be given a job. There was simply work to be done and people seemed to be happy to do it.

Other than getting tired from standing on your feet all day and needing to stretch now and then, cuttin' 'cots is easy work. Picking 'cots, on the other hand, is hard work. A 'cot picker is up and down a rickety 3-legged ladder all day and cannot avoid looking into the sun more than what is comfortable. Extending the arms out to pick fruit will build the muscles, but makes them quite sore first. Frequently the picker must reach into the tree and push the small branches aside to get to some of the fruit. The reward when withdrawing the arm is often not just the fruit, but a slap in the face from one of the small branches. The larger branches are scaly and rough, and tear at the skin. And splinters from the wooden ladders are unavoidable. But picking paid more than cutting, so the job was attractive to the older boys and young men.

Cuttin' 'cots, however, was different. It was a social event. It was "what was happening" for a lot of kids. Perhaps it was not on par with attending a high school football game, but at times it was close. In the cutting sheds, most of the kids were from the same area and went to the same school, so it was an opportunity to be with friends and earn some money for school clothes at the same time. The work stations were close together and, unlike picking, the work was light, so it was easy to talk to your neighbors or even to yell at someone across the shed. The radio was on, and it was possible to cut to *Lemon Tree* or *The Hop* equally well. When Ray Charles came on singing *Georgia on My Mind* it was like taking a break, because everybody joined in the call-and-answer of the song.

There were potential problems though. The cutting was done by the girls and the younger boys. The older boys were out picking. But sometimes an older boy, not a picker, would come into the shed and flirt with the girls. If there was not a man to run him off and the pickers drove up to the shed with a load of full boxes to unload and saw that going on, it could get ugly.

I was not a very good 'cot cutter. The entire process was awkward for me because I had not yet developed much dexterity in my hands. But I struggled through and was able to cut about four trays a day. That was a pitiful number, compared to what many other kids did, but that was no matter. To me it was a great success, and when the crop had been picked and cut, I was ready for further exploits. And closer to the privileged age of twelve.

21
The Old Farmer

After the 'cots were in, somehow I learned through the scuttlebutt that prunes were next, and flush with the success of cutting 'cots, I went looking for prune orchards. It was probably my mother who gave me the name and address of Mr. Pitman, who was a friend of a friend. So, one afternoon I hopped on my bicycle and peddled up to his place at the base of the Saratoga hills. There were other orchards on the way, but I don't remember stopping at any of them. I just rode straight up to Mr. Pitman's place.

Coming up from the highway, the driveway was steep for a little ways, and then leveled off. To the left of the driveway was an old, white, two story house encircled by large acacia trees. To the right was a well-kept orchard of old prunes trees. When I rode up, there in the driveway was the old farmer. He was standing with his hands resting on top of the handle of a shovel, a pose I was to see many times and which became one of my dearest memories of Mr. Pitman. He was in his early sixties, though his occupation had seen to it that he looked older. He wore old blue denim jeans and a very faded and threadbare long-sleeved shirt that was neatly rolled up to his elbows. On his head, covering his straight white hair, was a straw hat that was tattered beyond any I had seen before or have seen since. He was about six feet tall, of stout build, and his forearms rivaled those of Popeye—knobbed elbows and all. Those strong arms were covered with a pinkish, freckled skin, which was in turn covered with a thick layer of blondish hair. His hands were covered on the back side with the same blondish hair, and his fingers were as none I had ever seen. They were pictures of strength, sausage shaped, and bore testimony to the wear and tear of many years of hard manual labor. His face was that of an old Englishman who had spent a lot of time in the sun, and when he smiled the absence of teeth was revealed. His voice was a strong baritone and expressed a kind heart.

I met him there on the driveway and asked if he was Mr. Pitman, to which he replied, "Yes." I told him who I was and that I was looking for work. He told me that it would be a few weeks before the prune crop would be ready and advised me to come back at a certain time. Thus began a wonderful relationship between an old man and a young boy.

I will tell you one more thing about Mr. Pitman here. After my first year of picking prunes on the farm, I stayed on after the crop was in to help with some of the clean-up (described in a later chapter) and was stunned one morning as I met him. He had exchanged his tattered clothes for a new set of the same design. Even the hat had been replaced. In my eyes he might just as well have been dressed in formal attire, he sparkled so.

This was in the day when people had just recently begun to plant aluminum trees on the roofs of their houses. Those trees had a single, thin, brown root that traveled down the outside of the house, then entered into the home and terminated in a wooden box, out from which came images of every kind. Prior to the days of those trees and the boxes full of images, people, for the most part, bought what they could afford. But now, displayed before their eyes were all kinds of things that one could "easily" buy. The images were lures, both direct and indirect, enticing people to believe that one could have, indeed *should* have, so many things. Then along came that hard task master, Master Card, who gave people the means to fulfill sleeping desires that had now been fully awakened. And thus he enlisted a countless number to be his slaves.

But the old farmer was not among them. He bought what he needed when he could afford it.

And it was that simple.

All things are lawful for me, but all things are not helpful. All things are lawful for me, but I will not be brought under the power of any.
—Paul

Photo: Robert Freeman Pitman
at an unknown desert location

22
Picking Prunes

The first thing that should be said about picking prunes is that we picked prunes. We did not pick plums. In recent years the leaders in the industry, tired of their product being associated with the unpleasantness of constipation, have dropped the name "prune" in favor of the name "dried plum." Marketing folk do have a way of changing language to suit their purpose. So, "great" and "special" are used to describe common things, a government worker can be a "civil servant" or a "bureaucrat," and it is probably better not to call a person who is happy, "gay."

In fact, all prunes are plums, but not all plums are prunes. Plum varieties harvested for eating fresh are picked from the tree branches; prunes for drying are shaken out of the tree and gathered up. Unlike plums, prunes are also resistant to fermentation during the drying process.

Today, prune harvesting in the US is done with mechanical harvesters, but in the hey-day of prunes in the Santa Clara Valley they were picked up off of the ground by hand. A picker would be given two large galvanized buckets, which when filled would be emptied into a wooden fruit box. Two full buckets equaled one box. A shaker would go before the pickers with a long pole that had a metal hook on the end. The shaker would place the hook on a tree branch, carefully pull the branch toward him to take up the slack, and then shake the branch to release the fruit, which thundered down as hundreds of plump prunes hit the ground at nearly the same time. Shaking the prune trees had to be done carefully so that the branches were not broken. Sometimes, there was the unpleasant effect of getting a face-full of dust while performing this task, or of being covered with little white flies if a tree was infested. But it was, in my opinion, much easier than being on your hands and knees all day picking up the fallen fruit. That, dear reader, was *work*.

Each picker was assigned a row of trees in the orchard. There was no jumping around to find the best trees that had the thickest blanket of prunes under them. When you finished your row, you went back up to the top of the orchard and started on the next row that had not been picked. In the mid 1960s, the pay was 35 cents a box, and that included helping pick up the full boxes from the orchard, loading them onto the orchard truck, and then transferring them from the orchard truck to the truck that would be driven to the dehydrating facility. In earlier years, when the Pitman kids were still at home, the drying was done on site, and the rate per box also included helping with dipping the prunes[167] and setting them out to dry in the sun.

Good picking depended upon how large the prunes were, how thick they were on the ground, and of course upon one's own desire. Picking prunes was not nearly the social experience that cutting 'cots was. The trees were farther apart than the trays in a cutting shed and the rows were picked at different rates depending upon the yield of the individual trees and the skill of the pickers, so you were rarely abreast of the picker on the next row. Some kids picked a row together to enjoy each other's company, but almost all soon came to the conclusion that they needed their strength for breathing, not talking. It was also more efficient not to have to coordinate the effort of cleaning up under a tree, and talk seemed to slow the work down too much to make the reward for the effort expended worthwhile for most pickers. So prune-picking was largely a solitary experience, except when it was time to pick up the boxes and load the truck. Picking up the boxes was always welcomed. It was a chance to get off your knees, stretch your legs out and enjoy the company of the other pickers.

A prune picker needed to be outfitted properly. The prunes became ripe around the middle of August, so a broad hat was necessary to protect from the sun. Long-sleeved cotton shirts were worn because they protected from sunburn and kept the dirt out, yet allowed the skin to breathe. Denim pants insulated a little from the roughness of the dirt on which the picker knelt all day. Leather boots were best, but lacking these, high-top tennis shoes—made of canvas in those days—would do. Low-cut shoes made good collectors of tiny dirt clods and were useless. Speaking of dirt clods—knees don't like them. So kneepads of various kinds were tried by some of the pickers. Some pickers wrapped old rags around their knees and used tape to hold them up, others bought whatever kneepads were available at the hardware store. But due to the cost of good pads and the awkwardness of cheaper or home-made pads, most of the kids did without.

Prune fights on the Pitman farm were rare because wasting fruit was simply not allowed, and you just didn't do it. All the kids were instructed about this and the foreman kept a lid on it when Mr. Pitman was gone. One day, however, a few of the kids got into it, and some prunes ended up on the road and were left there. Mr. Pitman came by later, saw the prunes, and had no trouble identifying the guilty parties. After being severely scolded, I doubt that any of those kids ever threw another prune.

Picking prunes was a matter of overcoming. The first thing you needed to overcome was your bed in the morning. If you didn't start early, well before the sun was hot, it probably wasn't worthwhile to come out at all. That meant you had to go to bed early, and that required certain sacrifices. The next thing to overcome was the dirt clods. As soon as you knelt down to pick, you realized that there was some discomfort involved. There was also the stickiness of some of the prunes. The stickiness usually wasn't too bad, but when added to dirt it made a grimy paste on your fingers. Then, there were flies: big flies, little fruit flies, and tiny white flies. The fruit flies were the worst because they have a fascination for the inside of your ears and nose. Then there was the heat of the day. August was the hottest time of the year, and that's when the prunes were ripe, so you sweated profusely. Add all these together with a sore back from bending over and stiff legs from being on your knees all day, and you had a formula that sent some of the kids home before the first day was over. Others lasted a few days, and some—usually less than half of those who started—continued until the crop was in.

Now, you may well wonder why we didn't figure out a faster and less arduous way of picking prunes. Well, we tried. The idea was to use a large tarp and shake the prunes onto it. In those days tarps were made of canvas and were expensive, so purchasing one was out of the question. But my mother had a good number of old sheets, and I learned how to use a sewing machine. Put into practice, it quickly became evident that there was a fundamental flaw with that system. We could not waste fruit, and too many prunes had fallen on the ground before we came along with our tarp.

These prunes, and those that overshot the tarp during the shaking process, all had to be picked up. There was no efficient way to get them besides bending over from a standing position—and that got old real quick. It only took a few trees before the home-made tarp was retired and we went back to the way we had been taught.

You could easily see the struggle that went on in a new picker's mind. He wanted to earn the money, but he was not accustomed to the unpleasantness involved in prune picking. So overcoming was essential to success, and those who overcame developed the mental toughness of a marathon runner. A picker who achieved a twenty box day, or who could consistently pick more than fifteen boxes a day, was held in high esteem. He or she had overcome many things to be able to do that. I suppose that at a certain point a decision is made that goes something like this: "I'm already dirty and grimy. I want the money. I might as well just do it." And thus a life of discipline was learned by many young kids.

23
You Shouldn't Have Gone Along

In my first year on the farm, I was by far the youngest prune picker. Mr. Pitman never turned down a kid who wanted to work, so though I had just turned twelve, and most of the other boys were fifteen or sixteen, it was no problem to him.

But apparently it was a problem for at least one of the "young bucks," as Mr. Pitman would refer to us. I believe those kids were, for the most part, good kids, but it came into the mind of one of the older boys that a few of them should show their dominance over a relatively helpless subject. So one afternoon, when Mr. Pitman had gone to take a load of prunes to the dehydrator, four of these boys caught me when I was coming back from the restroom, and each grabbing a limb, decided they'd carry me out to the pasture and rough me up a little. I struggled to get free but my strength was no match for theirs. When I yelled, I promptly received a mouthful of dirt. I didn't really think they would hurt me and they didn't, but the experience was rather unpleasant and un-nerving. After about ten or fifteen minutes of dragging me around the pasture—which included through the sheep's business and over the salt lick—I suppose they had exhausted themselves and felt that they had proven their superiority, so they let me go.

I then faced a dilemma as to what to do, and I'm sure I stewed about it all evening. I was not willing to give up the job just because of some jerks, nor was I willing to tattle on them, not knowing what the outcome would be if I did. But I needed to do something to insure that the behavior of those boys did not continue. By morning I had devised what I thought was an effective plan. One of the boys had a car. It was an older car (kids who had newer cars didn't pick prunes), but it had a nice paint job on it, newer tan mohair upholstery, and it was no doubt the pride and joy of the boy who owned it. I had considered squishing some prunes and dropping them into the gas tank, but even at twelve years old and knowing almost nothing about cars, I had the sense that this punishment would be too severe for the crime. But Mr. Pitman had several kinds of fruit trees near the house and early the next morning when the other boys were out in the orchard, I picked up a few over-ripe

Elephant Heart Plums from off the ground. Elephant Heart plums have the deepest maroon flesh you can imagine, are quite juicy, and after laying on the ground for several days were perfect for making, shall we say, a design pattern, on the mohair upholstery.

Sometime later in the day, probably at lunchtime, the boy who owned the car discovered that his upholstery was now a kind of two-tone color. I'm afraid it broke his heart. He knew who had done it and why. When he saw me, he came up to me with shoulders slouched, and with the voice of one who had suffered a great loss said, "You didn't have to do *that*." I just looked at him, shrugged my shoulders and said, "You shouldn't have gone along." As I was walking away he said once more, "But you didn't have to do *that*." I replied again, "You shouldn't have gone along."

I felt sorry for that boy. He may have been the best of the bunch and proved to have a tender heart. But he was the one who had the car, he had participated, and I needed to make a point. The point was made, and they did not touch me again.

I've never understood that behavioral pattern, why some in senior positions, commonly for no reason other than sport, feel it is their obligation to pick on those not as strong as themselves. Even more, I don't understand why, unless there is a very strong base to the contrary, the behavior of a group of individuals frequently follows the lowest common denominator of character in that group. Years later, when in the military, I saw the phenomenon many times. As if pulled by some kind of degraded social gravity, good, seemingly upright young men sank to the level of the vilest among them. They just seemed powerless to do otherwise. They were sucked into a vortex of vice and profanity. The amazing thing is how quickly this condition can take over a social group. From fear of being the odd-man out, good people often fall into such behavior. So we commonly have hazing and worse—both verbal and physical—not only in the military and in fraternities, but also in other social groups as well. It is an amazing phenomenon. And one to guard against at all costs.

There are people who have collected extra letters after their names to whom we listen as they repeat what they have read or heard, who give some explanation for such behavior and offer advice on how to deal with it. But it seems to me that at best they only "heal the wound slightly." I'm afraid the infection itself is deeper.

I want to end this chapter on a positive note, so I'll tell you a story that is the antithesis of the story that I've just told you. It is the story of the Pitmans and of the family who pruned their fruit trees in the years just before the war.

In the Pitman family's first years on the farm, the pruning of the fruit trees was done by a Japanese-American family. They not only did a good job, but they were delightful people, and the Pitmans were quite fond of them. Then came Pearl Harbor. The Japanese family, like all those of Japanese ancestry living on the west coast, were required to go to an internment camp. Before they had to report for processing, they drove over to the farm and asked if they could store their furniture in the loft of the front barn. Mr. Pitman was more than happy for them to do so. From his youth, he had been taught to show respect to everyone and to be partial to no one. So, one day the furniture was put up in the loft. I think you can imagine the sad farewell.

Four years passed. Then one day the family came back, and all their possessions were there just as they had left them. If you know anything about the history of that period, you know that this transition for some of the Japanese and Japanese-American families did not go smoothly. Some families lost their homes, farms, and nearly everything else at the hands of people who had long before cast off all feeling for anyone other than themselves.[168]

When the Japanese family came back after the war, they acquired a small plot of ground on Prospect Road, near Miller Avenue, and grew strawberries. Their strawberry patch was on the way to the Sunsweet dehydrator, and Mr. Pitman would drive by during the prune season on his way there. It seems they would look for him, and on his return trip, they would flag him down and give him a crate of strawberries. Despite his strong objections, they would accept nothing for it and would insist that he take it. After a couple of times, he quit going that way, but rather drove a longer route to the dehydrator. He was quite uncomfortable that they felt that they needed to do anything for him. "It's just not right," he said, "It's just not right."

> *Whatever you want others to do for you, do so for them.*
> —Jesus

24
Clean-up

After the prune crop was in, there was still a fair amount of work to do, and though most of the boys took off, a kind young man named Dave Fowler and I stayed on to take care of that work. The props that had been used to support the tree branches needed to be picked up and stacked, the fruit boxes needed to be washed out and stored for the next season, and there were many other chores to do that could not be addressed during harvest time.

Dave and I picked up all the props out of the orchard, took them out to the back hill, then picked up all the fruit boxes and stacked them behind the dry-shed on the loading platform. During the course of the season, the prunes left a sugary residue in the boxes, so after lunch Mr. Pitman took me down to the dry-shed to wash them out. A very old machine was waiting for me there. It was a pump wagon about the size of a car that had four large steel wheels for transport and was entirely rust colored. Mounted above the wagon platform were two additional wheels, one about four feet in diameter, the other with a diameter of about one foot. The smaller wheel was fastened to the end of a motor shaft and a flat, fabric drive belt about four inches wide connected these wheels which were spaced perhaps eight feet apart. The photo at the end of chapter 7 shows a machine with similar wheels and drive belt.

For a boy who was raised in a housing tract, almost all of the farm equipment was strange, but the pump wagon took the grand prize in that department. Looking at the machine and the large stack of boxes was intimidating, but then Mr. Pitman told me to go into the dry-shed and strip down to my "skivvies." I came out—fair white skin in sparkling white underwear—to see and hear the monster fired up. The pump motor sounded like an endless string of firecrackers going off, and it wheezed as it gulped combustion air. The motor turned the little wheel, which pulled on the belt, which turned the big wheel, which compressed the water, which came out of a long, stout, black, rubber hose with great force. I was sternly told to stay away from the machine. No problem—with the little wheel turning fast and the belt dancing up and down making a whipping sound as it turned the big wheel—there was no desire to go near it. Next, I was given a demonstration of how to clean a wooden fruit box. If the sounds of the monster were not enough, the blast of the water that came out of the "business end" of the hose as it hit the wooden box completed the lesson in intimidation.

Mr. Pitman handed me the hose, which I immediately realized must be held very firmly. He then watched as I attempted to wash a couple of boxes. Not satisfied, he showed me again, told me that the pump could not be turned off, handed the hose back to me, then turned around and headed back up the road toward the house. I was thoroughly intimidated, standing there on the concrete loading area in my skivvies, with the monster roaring and water jetting out of the hose with a forcefulness I had never seen. The blast of water was so strong when it hit a box, that it was all I could do to hold onto it while it was being washed. The key was to keep one corner of the box anchored on the concrete slab while it was being washed. Even so, if the blast hit the box straight on, it would knock the box out of my hand. If the box was not held at the proper angle and I was not in the right position, the reward was a face full of water mixed with wood fibers, dirt and gunk. The first ten or fifteen minutes of washing boxes was interesting. The temptation to bail out and walk away was constant as I struggled to control the tail of the monster. But I gradually got the hang of it, and within about three hours there was a neat stack of cleaned boxes in the loading area. Mr. Pitman showed up not long afterward and turned off the monster. I was water-logged and my head was still clearing from the noise of the blast of washing boxes as I pedaled home that afternoon. But I had learned to work with the monster, and the next year it was easy.

After fruit trees have been harvested they are totally exhausted. They look wrung out, and in fact they are. For months they have vigorously pumped life-juice into thousands of offspring which are then severed from their branches. All the energy that went into producing the fruit has been expended, and there is nothing left. The weight of the fruit on the branches has been relieved, but that by violent shaking, and the branches and leaves clearly show the trauma that has taken place.

So the orchard had rest until the pruning in the late fall. And Dave and I went back to school.

25
The White Truck

The white truck deserves a separate chapter because it puts a lot of "science fiction" to shame. A true story is often stranger than a fictitious one, and perhaps the story of the white truck fits into that category.

The spring after my first summer on the farm, Mr. Pitman called me up to see if I wanted to help him during the spring vacation from school. I was glad to do so, in fact I don't know that I ever turned him down; and as I recall, I worked a few Saturdays for him during the school year also.

On one of those spring days, he sent me into the orchard to get something out of the white truck. I went out into the orchard, looked everywhere, but found no white truck. So I went back and told him that there wasn't a white truck in the orchard. This really hit his funny bone and he laughed a hearty laugh. Eventually, it became apparent that what was the white truck to him, was to me the old orchard truck we had used during the prune season. And it was anything but white. Okay, there was a little bit of white paint remaining on it. I suppose if you put all the little bits of white paint together, the surface area may have totaled a few square inches. But that truck was in fact *not* white, it was an *al naturale* variegated rust color.

Now, the white truck had indeed been white. It had been a Borden's milk truck in San Francisco prior to coming to the farm. But that was in the early thirties, it was then the early sixties, and a lot of water had passed under the bridge and over the truck in that time, giving the white truck a distinctive orange hue. The doors had been removed and the top had been cut off long ago. There was nothing that extended above the level of the dashboard. No roof, no windshield—nothing. The seat was a couple of old pieces of 2x6 lumber loosely held on by I don't know what. There was no back to the seat other than the protrusion of the bed of the truck that extended about three or four inches above the customized seat. The tires were bald, and at least one showed threads. That's threads, not treads. The truck had a steel bed with sides extending up about a foot and a half, and the tailgate was still on it. The hood, held on by bailing wire, was one of those that opened from the side. Each side folded up into the middle when in the open position.

The steering column was quite loose, and when you grabbed the wheel to hoist yourself up into the truck it came toward you about two inches. On the steering wheel, protruding from the middle of the shaft, just in back of the wheel itself, were two short levers about two and a half inches long. On the left was the spark adjustment, on the right the throttle. Yes, the throttle. Oh, there was a throttle button (not pedal) on the "floorboard" too. I put quotes around "floorboard" because in fact there was none, only the edges of truck body that the plywood floor-panels had once rested on. The plywood panels had gone the way of the doors, the top, and other unnecessary parts, and by looking straight down while driving one could see the ground being traveled over.

The brakes were of the mechanical type, meaning that they were not hydraulically assisted. Without hydraulics, or pressurized fluid, pure muscle power was required to operate them. Under the brake pedal and connected to it there was a horizontal shaft, and connected to that shaft were long, thin steel rods that connected to the back side of each wheel. Pushing on the brake pedal rotated the shaft, which in turn pulled the rods, which in turn expanded the brake shoes. The lack of hydraulics for the brakes was not a problem because the truck only went fast enough to use them when going down a hill, and that was always fun, so usually you didn't want to use them anyway. When not driving on a hill, the soft dirt of the orchard, combined with under-inflated tires and gunked up wheel bearings, slowed you down as soon as you backed off the throttle.

The clutch was also of the mechanical type, and pushing in the clutch pedal took a great deal of strength—especially for a thirteen year old. Since there was no back on the seat to push against, in order to get the leverage required, the driver needed to lock his left forearm in back of the steering wheel and pull against it. So, with all your strength you pulled against the steering wheel with your forearm, being sure to keep the steering column steady, tightened the muscles in your abdomen and pushed as hard as you could with your left leg. That's just to get the clutch in. You still had to shift the gears. The gear stick (of course there was no ball on the top) had had a spring under it and was at one time connected to the transmission housing by a pin. I said "had" and "at one time." By the time I arrived on the scene, both that spring and the pin that held the gear stick to the transmission had gone to wherever old truck parts go when they have finished their serviceable life. Gear shifting, then, was an art form in which the driver grabbed the side of the gear stick, pulled it up out of the gearbox, looked down into the hole in the top of the gearbox where the stick had been, located the exact placement of two parallel shafts with notches in them, carefully lowered the gear stick into one of the notches at just the right depth, then pushed or pulled the stick into the correct position for the desired gear. If any part of the procedure was not carried out correctly, there would not only be no changing of the gears but also an unpleasant audible complaint from the gearbox.

When going down a hill in the white truck, shifting gears was not to be attempted by a novice. It was better just to get the truck in neutral and coast down the hill, as I discovered one day. I was headed down to the dry-shed, and after opening the gate, started the truck down the hill. As speed increased, the backfiring of the truck became louder and I needed to get into second gear. But I missed my first stab at the notch for second gear, and as the truck picked up speed, the loudness of the backfiring was reaching a crescendo. Desperation set in, and the next thing I knew, I had the truck lurching harder and harder against the center of a fence post and backfiring louder and louder every second. Mr. Pitman came running down to where I was and reached in to turn off the ignition. I was quite embarrassed and tried to explain that I had missed the shift, but he just banged on the steering wheel and exclaimed, "*This* is the important part!"

Oh.

Now, dear reader, lest you get the wrong idea and think that this is in any way a complaint, you must understand that the white truck was a great friend of mine and of some other young kids, on two counts. First, you got to drive; and for a thirteen year-old, that was pretty good stuff. Second, hauling things with the truck was a lot easier than carrying them on your shoulder.

So the white truck took a while to master, but it was the means by which several young kids learned to drive.

A satisfied soul loathes the honeycomb, but to a hungry soul every bitter thing is sweet.
—Solomon

A truck very similar to the "white" truck. And nearly in as good condition.

26
Harmony

Mr. Pitman was a wonderful teacher. He taught by example, not so much by words. His farm was as neat and orderly as any I have ever seen, and yet I never remember him saying one word about neatness. He simply set a standard, and if you worked for him, you kept that standard. Tools were put back in their proper place, boxes or other items were stacked neatly, the dry-shed, garage and tank house were kept clean, and there was never a bit of litter that remained after it was seen.

One of the highlights of working on the farm was getting to work with Mr. Pitman. For me, that did not happen all that often as he was wearing out by that time, but when it did, it was a special treat. One time in my first years there, before I became stronger, he and I had struggled with a strenuous task for some time. After a while, he eased back, wiped his brow and said, "Between an old man and a young boy this thing has a pretty good chance."

His talk, though about common things, was not common. It felt good going down. There was wisdom in his words, and they were seasoned as if with salt. He did not boast, nor complain, nor talk about foolish things, and his expression was that which comes from a kind and loving heart.

One of his sayings was, "You can't have a good day until you perspire." For him, to perspire was akin to having his blood flow. Sweating was a flow of his life juice, and perhaps because he drank fresh well water, ate fresh food, and perspired heavily every day, I don't remember any unpleasantness associated with it. Of course, we were usually outdoors where the living air circulated through the freshness of the trees, the grasses, and the earth.

Another of his expressions, which he used sometimes after assigning a task, was, "Have fun." And he meant it. It didn't matter what the task was.

Mr. Pitman was a peaceful man, a man with an inner compass. In all the time I knew him, only once did I see him lose his peace. One morning when I met him he was just stewed up and said something about "kids smoking pop" [sic]. He must have seen something on television, and it really upset him. His frame of mind was quite out of character for a man so at peace, and it puzzled me. He did not retain the condition for long though, and it was not until decades later, when writing this book, that I understood. I did not know at the time I worked for him that he had spent most of his life working with kids. He invested heavily in young people and cared greatly for all of us. His son, Dave, told me about this many years later. Dave also told me how much his dad enjoyed working with the kids who came to work for him. That surprised me. I suppose I had never thought about it from his side—I just felt that I was the fortunate one. I think that on that day it just hurt him deeply to see kids being wasted by drugs.

I never heard Mr. Pitman cuss, though I saw him struggle in plenty of situations in which some would think profanity most appropriate. Farm work has its share of frustrating tasks. On occasion he would mutter something about "Sam Hill," but it was never with anger, and at the time I had no idea what he was referring to. I greatly appreciated this tenderness toward me in later years when working with some other people who did not possess the same graciousness. There was just something so right about being with him that it made you understand that it was a privilege.

Mr. Pitman was not a joking man and had no need to be entertained. He was simply content. And that was really something special to be around.

I'll tell you about his vegetable garden here. I never worked in his vegetable garden. It was his, and he obviously loved it. The garden was in back of the house, outside of the ring of trees that encircled the house. There was only one gate to the back acreage, and coming up from there, you had to pass through that gate and by the garden. Much of my work was in the back, and many times in the cool of the late afternoon, when I headed up toward the front to go home, I would see him in his garden on his hands and knees working. Sometimes he would see me, and we would greet each other, and sometimes he would send some squash, or tomatoes, or something else home with me. But I never interrupted him there. I may have sensed that it was sacred time. And it probably was. I am certain that he had his share of unpleasantness to bear, and his humility and kindness came from somewhere. Looking back on it now, I don't think that he was just pulling weeds or picking vegetables when in the garden.

He knew the secret of keeping weeds out of a garden—pull them while they're small, before they have a chance to go to seed. After a while you just won't have many weeds.

Catch us the foxes, the little foxes that spoil the vines.
—The Song of Songs, which is Solomon's

27
The Bucksaw and the Man-Killer

Everyone is afraid of something. I knew a man who seemed to have no fear of anything other than *not* breaking bones or otherwise getting all beat up on a regular basis. Without this he had nothing to boast of. An odd case maybe, but the point is that everyone is afraid of something.[169]

For me, on the farm, it was the pump wagon that we used to wash the fruit boxes out with, which I've already told you about. For Dave Pitman it was the bucksaw. A bucksaw was a big circular saw blade driven by an electric motor or gasoline engine. They were used before modern, lightweight chainsaws came on the scene, mostly for cutting up trees for firewood; and nearly every farmer had one. The blade was usually about 24 to 30 inches in diameter, and was bolted to a shaft that had a pulley on the other end. That pulley was connected to the motor by a drive belt, which was like the fan-belt you have under the hood of your car, just much wider, flatter and longer. Bucksaws were mounted on the simplest of wooden frames with no table, just a few boards around the blade for you to rest a tree limb on. When a bucksaw was turned on, it was downright scary. There was no guard for the blade. There was no guard for the belt. And when working around one, you had better be thinking clearly and keep good footing.

Bucksaws are the kind of device that younger people today find hard to believe ever existed. At that time there was no Occupational Safety and Health Administration. Nor were there five or six warning tags on an electrical extension cord that you purchased at the hardware store. The bucksaws we used clearly displayed the absence of OSHA, and in some respects may have been safer to operate than some saws with fourteen warning labels and safety devices on them. At least it was glaringly obvious who was responsible for your safety, and that you had better have a proper respect for the tool you were using. It seems that at that time no one ever thought of any kind of safety device other than using one's head. But sometimes that wasn't enough. One hundred and thirty-seven volumes of OSHA regulations later, sometimes it still isn't.

There was another device on the farm that was as fearsome as the bucksaw, but in a different way. It was known on the farm as, "The Man-Killer." It really was just a heavy, straight wrecking bar about five feet long that weighed about twenty pounds. It differed from "The Boy-Killer" only in that it was larger and heavier. If something needed to be torn apart, or if there was a hole that needed to be dug through sandstone, there was nothing like it to get the job done. But lifting it repeatedly… well, that's how it got its name.

A wrecking bar is also commonly used as a lever over a fulcrum block. The principle of a fulcrum is fascinating. Someone once told me, "Give me a long enough bar and I can move the world." Well, almost. The fulcrum is one of the best friends of the farmer or construction worker if he understands it adequately. The use of a lever over a pivot block enables you to lift or move a mass proportional to the length of the lever and the position of the pivot. There are virtually no limits if the correct proportions are used. A lever is not a tool of ignorance. As a matter of fact, I think "unskilled or "skilled" labor should be defined by the noodle applied, not by the task performed. By thinking out a move, many farmers and construction workers have lifted beams or stones of thousands of pounds into place without the aid of a helper other than a lever and a pivot block. There is also great satisfaction in overcoming the effects of gravity on mass by using your head.

One Pitman farm alumnus tells the story that several years after leaving the farm, when he was working on a construction site, the foreman came up to him one day and said, "I like to hire people like you—people too lazy to work hard. You have to make a living just like everybody else, but you're too lazy to work like everybody else, so you'll always come up with a faster, easier way to do things." After he got over the shock of the foreman's bluntness, he took it as a compliment. The ability to do that he said, he owes to the old farmer. And to the Man-Killer.

A Bucksaw

28
Other Work

A farm does not keep itself up. The way a farm looks and functions is a direct result of plain old-fashioned labor properly timed and executed.

On any farm, there is always shovel work to do. And a lot of it. After all, farming is about dirt. If a farmer is not turning over the soil with a shovel, or planting with a shovel, or controlling irrigation with a shovel, he is digging a trench, or digging a hole to bury garbage with a shovel. So a shovel is an extension of the farmer's hand.

There are two kinds of digging with a shovel: the necessary digging and the extra-curricular digging—digging not required, but done because it accomplishes the task in a better way. Mr. Pitman preferred the extra-curricular type of digging. I mean, the man *liked* to dig. The electrical and water lines from the house down to the dry-shed were put underground. No back hoe or other tractor was used; it was all hand dug. The trenches for the irrigation pipe for the pasture sprinklers were dug by hand. There were also underground water lines to two large hydrant-like water terminals out in the back. Yep, all hand dug.

In the earlier days of the farm, Mr. Pitman did most of the digging himself, but in later years, others, yours truly included, earned a degree in the art. And an art it is too. If you are going to last all day working with a shovel, you need to learn how to work with your tool and the dirt and not fight them. There is a rhythm to be found, and once in that rhythm it's not so bad. But first, the skin on your hands needs to get toughened up. Tender skin will blister easily, but after enough blisters, the skin toughens and blisters are rare.

Besides gaining a great respect for shovel work, I also learned many things while using that tool that can make a fool out of one who does not understand it. Once, when digging a ditch across the natural drainage for a new water line, I ran into some clay that was nearly as pure as that which we all played with in school when we were younger. The clay was fascinating to me, and I studied it thoroughly. I now knew where clay came from. I mean, I really knew. There is a difference between knowing and *knowing.* But the clay made the digging miserable.

Wild roses grew near the drainage where I was digging that ditch, and while there I gained the fondest appreciation for them. The contrast was indescribably beautiful. There I was occupied with a most arduous digging job in the clay, yet surrounded by the tender beauty of wild roses. It was like hearing Louis Armstrong, with his gravely voice, sing *It's a Wonderful World.* Every time I see wild roses I remember that time in the ditch.

There was also some work on the farm that made ordinary shovel work look like fun. A lot of farm work is quite rewarding. Looking at the boxes of prunes stacked up at the end of the day, or seeing the orchard all cleaned up after the harvest, or seeing a completed ditch—all that was rewarding. Picking rocks was not.

Rocks in an orchard, field, or in most any planting area are very undesirable. Rocks dull the disks, and that makes the task of disking more difficult. If that is not bad enough, the unpleasant sound of disk to rock contact is akin to that of running a fingernail across a chalkboard. Anyway, the rocks have to go. By the time I got to the farm, most of the rock-picking work was already done, and the road down to the dry-shed was paved with the proceeds nearly the entire way. But there was one lower section of orchard, where Paramount Drive now joins the highway, where there were still some rocks. In Mr. Pitman's mind, these rocks had my name on them. Now, according to Dave Pitman, Mr. Pitman's younger son, picking rocks was not so bad because, in his words, "That's when I got to drive the truck." But for me, I was already driving the truck when this prospecting work was first introduced, so I had "less fun" picking rocks. It was not until many years later that I realized where all the beautiful old rock walls came from that were at the edges of many of the farms in the hills. Every orchard needed the rocks picked out of it, and a *lot* of people picked a *lot* of rocks. When you drive through the area today it is difficult to find any of those old rock walls. They are part of a history that has been largely covered over. But I do wonder if the homeowners on Paramount Drive, where the road down to the dry-shed used to be, ever wonder where all those rocks in their yards came from!

There were also the reeds, or spike rushes, in the pasture. The spike rushes grew in the drainage and were taking up precious pasture land. They grow in clumps and have a root system that is thick, deep, and matted like a gigantic Brillo pad. One summer, Mr. Pitman declared war on them and drove the tractor with the disk over them for a few days. All he got for his effort was a bumpy ride and an empty tank of diesel fuel. The reeds as much as laughed at the disk. So after he had had it, the job was handed over to me to do by hand with a mattock. A mattock is like a pick, but has a flat edge instead of a pointed one. I was shown how to dig one of the plants out and was promptly left alone for a week while Mr. and Mrs. Pitman went on a vacation. It was probably the most miserable work I ever did. Except for shoveling manure.

Frank Izadore was a cattle dealer and a friend of Mr. Pitman. Mr. Izadore had the ability to look at an animal and tell almost exactly what the animal weighed. He went around to the farms and bought animals, which he re-sold, and so made his living. But times had changed, his small stock yard was now surrounded by housing tracts, and the neighbors were not fond of the flies associated with the cattle business.

Occasionally Mr. Pitman would go over to Mr. Izadore's to get a load of manure. This was mutually beneficial—Mr. Izadore's fly problem was lessened, and Mr. Pitman got a load of good fertilizer. Every spring Mr. Pitman put a thick layer of manure down in his garden and disked it in. Though I never tried it, I don't doubt that one could have planted a broomstick in that garden and gotten it to grow.

One afternoon after I had completed my chores in the back, I was greeted by a load of manure and a pitchfork. The prospect was at least a little repulsive, but I reluctantly got up on the load and started pitching it off. The problem was that with each fork full of manure, a wetter under-layer was exposed to the air and the heightened aroma was just too much for my tender nostrils. I picked at the manure for a while, and though I loved the old farmer and didn't want to disappoint him, I had to go tell him I just couldn't do it. That was hard to do, because he never asked me or anyone else to do anything that he did not do himself. I felt bad that he got stuck with the job, but the contents of my stomach were much happier to have it that way.

Some of the work done on the farm outlasted the farm for decades. Shortly before I started working for Mr. Pitman, he planted three rows of pine and cypress trees around most of the property. With the widening of the highway (Saratoga-Sunnyvale Rd.) he lost a row of walnut trees, and I suppose he planted the trees between the highway and the house to serve as a buffer from the ever-increasing traffic noise. Why he planted the trees around the back acreage, I don't know. Perhaps he was an "environmentalist" before the term came along. Those trees were very important to him and a great deal of effort was required to grow them from seedlings to self-sufficiency. That became my charge. For four years I watered the trees and kept them from being choked by the weeds. A lot of shovel work went into digging and maintaining a "pot," or circular ditch, around each tree to hold the water.

Not many of those trees remain today. I suppose they got too big. But for years afterwards, I'd drive by now and then to check on my trees.

So the tasks were many and varied. And then there was the sheep…

29
Sheep

Sheep had been part of the Pitman farm from the beginning. Their pasture was in back of the prune orchard, to the north of the apricot orchard, and included the slope and bottom land down to near the edge of the drying area where the fruit was set out. The pasture had a substantial fence around it, had its own sprinkler system which enabled the grass to grow all year, and had three large cast iron kettles for watering troughs.

The sheep were Suffolk and Hampshires—beautiful breeds—and Mr. Pitman kept thirty ewes and one ram. In the spring, when the lambs were born, most of the ewes had twins.

The beauty and delightfulness of the new lambs would be hard to exaggerate. In my four springs at the farm I never saw a birth, but I did see several lambs shortly after birth that were still wet and being cleaned off by their mothers. The tenderness of the scene is marvelous in the true sense of the word. Within hours the lamb has, with considerable struggle, gotten to its feet. Though very wobbly to begin with, within a few days those skinny little lambs have become the most agile athletes imaginable. Observing the jumping and skipping of the newborn lambs could melt a heart of stone. In their frolicking, the pure joy and goodness of life is displayed in all its grandeur. I count it one of my great privileges to have been allowed to watch the lambs during those years.

But all life on earth is fragile, and one day the harsh reality of that fact was sadly shown to me. I was headed down to the dry-shed early one morning to begin work, and there in one of the water troughs was a little drowned lamb. Though the troughs stood about sixteen inches high, the lamb's center of gravity obviously got higher than that. With a heavy heart I buried the little fella.

As the lambs get older and stockier they gradually lose their nimbleness. Also, the gentle, peaceful nursing of the young ones gradually gives way to more forceful practices. The weaning of the lambs can be a demonstration of innocence lost. When they become larger and hungrier, they learn that if mom's udder is struck hard, more milk is discharged. So instead of sucking gently, a pushing of the head into mom is practiced. At first, the pushing is relatively moderate, but in time it can develop into just plain bad manners. Of course these thrusts into her body are not comfortable for mom, and when they get too rough she starts walking off. If the little ones do not want to get their heads crushed by her thigh, they back off.

Some of the lambs find the transition to earning their own living by eating grass quite distasteful. They'd rather just live off of mom as they had when they were younger. The problem is that mom is pretty well worn out from the whole process of child rearing and especially from being bumped so hard. She allows them to nurse less and less, stares them down when necessary, and eventually they get the picture. Some will sneak in for a quick thrust now and then, but soon realize that that is unprofitable. So it is with most. But with some individuals, weaning becomes nearly a war. Certain youngsters just don't care about their mom. I suppose they hold the opinion that her udder is their God-given property and they will take what is rightfully theirs—and do so forcefully if necessary. You can actually observe them scheming, trying to get on mom's blind side and strike when she's not paying attention. Mom wheels around and dispatches a head butt, which if well connected, begins to teach the lamb that perhaps such behavior is not acceptable. These battles can turn downright ugly when a young one becomes defiant and mom has to drive him away repeatedly. But eventually all the lambs are eating grass.

A little later in the spring, after the ewes had recovered from birthing, it was shearing time. All the sheep were rounded up and placed in a holding pen just in back of the garage. Chipper, the sheepdog, took the lead in the roundups and these were, without doubt, his proudest moments.

The end of a shepherd's staff has a thick "R" shaped doubled wire on the end. The bottom of the "R" is a wide opening for hooking the sheep's leg and the narrow neck is for securing it. Mr. Pitman and I would go into the pen, herd them into a corner and I, being quicker than he, would use the staff to grab the rear leg of one of the sheep. I would reel the animal in and we would flip it over on its back. I'd grab the rear legs, he'd grab the front ones, and we would carry our prize into the wide back door of the garage to be sheared. All the sheep's strength is in the rear legs—they have no power in the front ones—so I was warned sternly to take care of "the business end." I never got kicked, but I've heard that if it happens to you once, you will make sure it never does again.

The shearing process was fascinating. I kept hold of the rear legs while Mr. Pitman used an electric wool clipper that was similar to a barber's clipper, just a little bigger. He started with a run up the middle of the belly, and working quickly and keeping the fleece all in one piece, worked to one side as far as he could go. Then I'd lean the sheep in the opposite direction and he'd do the other side. The dangerous part was finishing the last few stripes down the back and detaching the fleece. By that time the sheep was tired of the whole process, and being nearly upright, was in a position of strength and knew it. But I always held on, though sometimes with considerable effort. Visions of a half shorn sheep running around with fleece attached kept me alert. Thankfully I never had an escapee, although I'm sure some other shearers have. The shearing done, Mr. Pitman would powder the animal with tick powder and then I'd let it go directly into the pasture—bleating, pink-naked and embarrassed, but unhurt.

It is impossible to adequately describe the beauty and fragrance of a freshly sheared lamb's fleece. The animal may look quite dirty on the outside, but the inside of the fleece is as white as snow.[170] It is absolutely beautiful and the feel of it is heavenly. It is not only soft and springy, but also slightly oily and perfumed. As the animal gets older, the quality of the wool diminishes. A lamb's fleece may be about eight inches thick, sparkling white, and as soft as soft can be. They also rarely had even one tick on them. However, by the time a sheep is several years old, the wool is coarse and may be only about three inches thick or even less. The pleasant aroma is also gone and they are far more susceptible to ticks. And mutton, by the way, is quite inferior to lamb.

I would roll the fleece up into a tight ball and Mr. Pitman would tie twine around it to keep it together. Then I would climb a ladder and drop the fleece into a long gunny sack that was hung from the rafters. As more fleece were added to the bag, I would jump into the bag to compress the wool. It is amazing how tight wool compresses, and as I recall, we usually got all the wool into one bag.

When the sheep were all shorn, it was time to take care of the lambs. Sheep are born with tails, but tails are undesirable when it comes to wool production. Likewise, male lambs that are raised for meat are castrated. So after all the lambs had been born, we would have a round-up to "band" all the lambs. We herded them into the holding pen, caught them, and placed tight rubber bands around their tails and the scrotum of the males to choke off the blood to these appendages. As I held the lamb on his or her back, Mr. Pitman loaded the banding instrument. He would ask me, "One or two?" And so was my first sex education class. When the banding was completed, all the sheep were let out to pasture, the little ones complaining a bit. A couple of weeks later, lambs tails would be seen lying here and there in the pasture grass.

Although sheep are beautiful and docile animals, they are not very intelligent. It is one thing to be told how dumb sheep are; it is quite another to observe it day after day, year after year. A remarkable example is the following instincts of sheep. Very often, and for no apparent reason, they just start following one of their number who has taken but a few steps in a certain direction. A line forms, and it does not matter if it is a lamb at the front or an elder. With the occasional exception of a very old ewe, when one goes, they all go. Sometimes the one in the lead realizes they are leading and stops, so the line breaks up and that's the end of that. Sheep can also easily become lost in familiar territory and can become frightened, even by the wind. But for the classic case of the mental limitations of these wonderful animals, I'll first need to give a little background.

By the time I came to the farm, Mr. Pitman had given up on the apricot orchard in the back. It was a lot of work to keep it up, the kids were grown and gone, and the orchard was really too small to be profitable. So, nearly the entire back portion of the remaining farm had become an extension of the sheep's domain. (The "back" acreage is the portion of the farm shown in the photo at the end of chapter 16.) There was only one opening connecting the pasture to this additional grazing land, and that was a side gate at the top of the hill. The side gate was nearly always open, and during the course of a typical day the sheep would wander from one area to the other. With some frequency, a nursing lamb would become separated from its mother so that one of them would be on one side of the pasture fence, and the other on the other side. Since nursing through the fence was impossible, they would begin to cry for each other, not knowing how to come together, though their distance from the side gate was usually less than a hundred yards. The little lamb crying and not knowing its way we can understand, but the ewe, who had been there for many years and who had been in and out of that gate countless times… well, that's a little harder to understand.

When working in the back acreage, I would often hear the unmistakable crying of a ewe separated from her nursing lamb, and soon learned that if I did not want to listen to it all afternoon, I'd need to help them get together. In fact, they might never figure it out, so when I could, I would go down and chase the one who was on the grazing area side up the road toward the pasture gate. The ewes, after getting onto the road, or up it a little ways, would know their way to the gate and would run full speed up the hill to it, bleating for joy the whole way. More than once, I found both ewe and lamb lying down on opposite sides of the fence exhausted and hoarse from crying. And to think that in the Bible we humans are repeatedly referred to by God as sheep!

I would like to think that I am smarter than that. But frankly the evidence is to the contrary.

We all like sheep have gone astray.
—Isaiah

30
Chipper

In the last years of the farm, Mr. Pitman had a dog named Chipper. Chipper was a black and white sheepdog, and a watchdog extraordinaire. He also had a mind of his own.

As far as Chipper was concerned, he was king of the domain, the Pitmans were his chief servants, and all who came onto the property were his subjects. Any car that came up the driveway was greeted with barking and an incessant biting at the front tires. Some folks, not accustomed to such behavior, would stop their car thinking that they might hit the dog. Someone would then have to come out to them and explain that that was not possible. They rarely, if ever, believed what they were told, and were thoroughly intimidated by the dog. And I'm sure Chipper wanted it that way. He had to let you know that while you were there, you would be subject to him. Chipper wasn't mean and didn't snarl unless provoked, but he wanted you to know that the farm was not your place; it was his. When Mr. or Mrs. Pitman came out to see who had arrived, Chipper would proudly present his hostages, and having handed them over to his underlings, would go and carefully place a valet ticket on one wheel of their car so that all would know who it belonged to. He would then strut off to go check on the rest of his domain. It did not matter how many times someone came, even if it was Dave, the Pitman's son. Still the tires must be bitten at, though sometimes the valet ticket was good for multiple entries.

The last four years that I was there, which ended up being the last four years of the farm's operation, I supplied the prune-picking crews. The kids I hired were all friends from school, and these kids were not of an age to have cars. They had bikes. But the greeting procedure was essentially the same. I warned all of the kids ahead of time what to expect, and told them to just keep peddling up to the top of the driveway. I made it clear that the dog was not really a problem and that he would not hurt them or their bicycles. But when it came down to it, when they rode up the driveway and were greeted by Herr Chipper, almost invariably they stopped. After all, the dog was biting at their tires. To my knowledge he never punctured even a single bicycle tire. His biting was just a kind of accompaniment to the barking, but these kids had a difficult time grasping that concept when all appeared to the contrary.

The problem with stopping in the driveway was that Chipper would then immediately issue a valet ticket. And sometimes not just one. If you resisted or complained in an improper manner, he might just empty his entire ticket book. I made many trips out to the driveway to find one of my school chums looking quite chagrinned, having had his precious bicycle duly peed on. The fact that these kids stayed on to go to work there says something about their desire.

Anyway, as long as you behaved yourself and did not challenge His Majesty, the ticket was good for subsequent days and the procedure of checking your bike did not have to be repeated. So really, all was well as long as you were checked in properly.

The prune picking season usually only lasted about three or four weeks, so unless Mr. Pitman had a special project, it was just he and I working there most of the summer. So I got to know Chipper a little more intimately. I tried with considerable effort to befriend him, and at times he would act as though he had accepted me and we were friends. But in fact it was not so. I suppose it was just his breeding. Besides, it was his place. And we were not equals.

Chipper had a unique way of informing me of his status. The idea was to catch me in a confined space and apply a jaw clamp to the top of my boot. He knew exactly how hard to clamp down. He never hurt me—that wasn't the point—but it sure was an inconvenience. And that *was* the point. I did not have free run of the place, he did. So he chose his place and times to remind me of that fact. The tools were kept in a side room of the large garage (also known as the upper barn), and the door to that side room was mid-way into the garage. The Pitman car was usually parked in the space closest to the tool room, so when Chipper knew that I was headed there, he'd crawl under the car and lay in wait for his victim. In I would walk, out he would lunge, and on went the clamp. If I tried to resist, he would tightened the clamp. If I did nothing, he would relax a little, but not let go. And he would not be persuaded. Sweet talk, petting, and other social graces were useless. He let go when he wanted to. It was time for him to instruct me, and it was not time for me to do anything but submit.

Well, I didn't have time for that kind of nonsense, and was not about to put up with it, so it became a little battle of the wills and skills. I worked quite hard at it and I'm sure he did too. He was intelligent enough to never attempt the jaw-clamp when I had a tool in my hand, and I was not intelligent enough to always remember to have one in my hand when going into the tool room, so I suppose we were pretty evenly matched. When faced with the challenge of the grip on my boot and no tool in hand, my goal was to raise my foot quickly with his jaw still attached and clang his head on the undercarriage of the car. I sometimes succeeded, but not that often. If score was being kept—and it *was* being kept—I was losing. So over time I changed my tactic, and being a little stronger than when these skirmishes first began, I began grabbing him by the skin on the back of the neck and putting the clamp on him as he had on me. This was met with great displeasure, as I was now the one in control. So the conflict escalated a little and he became bolder. One day, as I was on my way to the tool room, Chipper decided to put the clamp on me out in the open. That was not a smart move. I grabbed him by the scruff of the neck with both hands and in one motion picked him up, whirled around and threw him up against the pasture fence. Something clicked in his understanding at that point and he backed off and rarely troubled me after that.

Occasionally Chipper would still make a feeble attempt to reclaim his lost dominance, though he changed his spot to the narrow passage between the house and the tank house. I suppose he hoped that one of his servants who lived in the house might come to his aid. But they never did, and by that time I was too big and strong, so when the clamp went on I just kept walking, dragging the dog that was attached to my boot along with me.

In spite of this behavior, Chipper was, at times, fun to watch. He'd get something in his head—suddenly realize that he needed to check on something somewhere on the farm—and off he would go to his task. Whatever those tasks were I never could figure out, but they were important to him.

I had my work; he had his.

This dog reminds me of Chipper

31
The Whizzin' Machine

The farmer's life is not an easy one. There are a very wide variety of tasks that need to be done and resources are limited, so ingenuity and hard work must substitute for cash. Many tasks can be approached in different ways and yield the same results. With the volume of work to be done and limitations on strength, time and resources, it is essential to success that one learn not only to solve problems, but to do so efficiently and without spending money.

When I was working with Mr. Pitman, he would occasionally laugh at himself and say, "stupid farmer." Actually, he was anything but stupid. He was an expert at making do with what he had and finding solutions to difficult problems.

The "whizzin' machine" is a good example. I've mentioned the old orchard truck. His other truck, during my first years there, was not all that much better. It was also a Model AA Ford truck, and though in much better shape than the "white truck," the yellow truck, as it was called, was also well past its prime. Now and then, one of the trucks or the car would get a flat tire, and in those days that meant taking the wheel off, removing the tire from the wheel (which is no small feat when done by hand), and patching an inner tube. Then the tire needed to be put back on the rim and inflated. Unlike today, manufactured goods were expensive fifty years ago. For Mr. Pitman to purchase an air compressor for the purpose of inflating tires was out of the question. But he salvaged a motor and compressor off of an old refrigerator, then purchased a length of narrow rubber hose and a nozzle tip from the hardware store. He then built a wooden dolly to mount the motor and compressor on, and Viola! the Whizzin' Machine! The fact that the little compressor took an hour to fill the tire with air was no problem at all—we just started it before lunch and disconnected it after lunch.

Another example of the old farmer's ingenuity was the "gin pole." Need to lift something real heavy? No problem, get out the gin pole. The gin pole was just a wooden pole, that is, a straight tree with the branches cut off. If the lift required a tripod, three poles were used and were tied together at the top (see photo at the end of chapter 7). Up in the loft of the dry-shed, Mr. Pitman kept hundreds of feet of rope and several pulleys, or what is commonly called "block and tackle." Now this was not just any block and tackle. The "blocks" containing the pulleys were large and of very old design. They were made of oak with a steel hook and wheels, like those in the photo at the end of this chapter. There were blocks with one, two or three pulleys. The old hemp rope was very thick, perhaps an inch in diameter. All the equipment was of the same vintage. If you have ever seen a replica of a 19th century sailing ship, it probably had rigging of the same design. It is likely that an old ship was the source of the equipment we used.

Block and tackle go back a long ways. Ever wonder how the old-time sailors had the strength to move those huge sails? They had Archimedes of Syracuse (287 B.C. - 212 B.C.) to thank. Archimedes was a Greek mathematician, physicist, engineer, inventor and the like. He was a really smart guy who invented practical stuff, like the block and tackle pulley assembly. How he figured out that he could gain a mechanical advantage by running a rope though two or more pulleys, I don't know. And the math he used to describe the increased advantage of additional pulleys, I can't follow very well. But he did figure it out, and with his help we lifted the well house up and off of the well, and others circumnavigated the globe in sailing ships. I still don't understand why the system works, but it is easy to gain a four-fold lifting advantage using just two blocks. One block, with two pulleys, is fixed at the top (of the gin pole on our case); the other is attached to, and moves with the load. There needs to be a certain distance between the blocks for maximum advantage, and with a little experience, that distance is easy to find. If a higher mechanical advantage is needed, blocks with more pulleys are used. By looping the rope through more pulleys, and placing the blocks properly, a proportionally higher advantage is achieved. The ease with which heavy objects can be moved with block and tackle is astounding. The load is controlled and secured by use of the ropes and an equally ingenious system of knot tying.

I've mentioned that Mr. Pitman was a good teacher. But his methods, at least with me, were not with a textbook, nor even with much verbal communication. He simply set an example, and you followed it. When coupled with necessity, his teaching method proved very effective. Once, after I had worked with him for a few years, I was picking up the props in the prune orchard after the crop was in. I had loaded the "white" truck with props and was ready to take a load out to the back hill. But the truck wouldn't start, so I walked over to the house to tell him. He stepped out, and I told him the problem. "Well, I guess you'll just have to carry them back there." That was it. Nothing more. He turned around and went back into the house.

Well, I certainly wasn't going to carry all those props all the way out to the back hill, so I went out to the truck and opened the hood to have a look. Now, I knew nearly nothing about how a truck engine works, and he had never taught me anything about it, but I could clearly see the fuel sediment bowl filled with gasoline and a tube that went down from it to the carburetor. So I tapped on the sediment bowl and fuel line with a stick, played around with the carburetor a little—pulling the levers and moving the flapper valve; wiggled some wires that I thought might be important, and got in and started the truck. I have no idea what I did, if anything, to enable the truck to start, but I learned a big lesson. This was also something that could be mastered, and I was no longer intimated by it.

There were many such lessons in problem solving. And of course I learned, as all farmers do, that nearly anything can be fixed with bailing wire and a pair of pliers.

I owe much of my ability to make a living to Robert Pitman, and I'm sure there are some others who worked for him who would say the same thing. After a few years of working with him, you realized that there was always a solution and that a given task could almost always be done with whatever was at hand. A friend of his put it so well when he said, "He taught me how to think."

And many of us would agree.

Photo: Block and tackle

32
The Help

Farm labor in the Santa Clara Valley, and in all of California, has been a challenge since the mission days. By nature farm work is seasonal. No matter what combination of crops you plant, there are always going to be down times for most farm workers. There are simply more workers needed during the harvests than there are at other times of the year. Even at the end of the agricultural era in the Valley, when efficiency was at its height, farm and cannery workers were out of work for at least a few months of the year. The question for the farmer is: "Where can I find people who will help me for part of the year and not be a burden to me for the rest of the year?" The answer for most of the farmers during the fruit growing era in the Valley was: right at home and next door.

I mentioned early in this book that the orchard farms in the Santa Clara Valley were mostly family farms. The families worked them—the whole family. For the most part, labor was not something foreign, done by outsiders; it was Dad and Mom, the kids, and the kids' friends from school.

The work of children in the orchards is a big part of the history of the Santa Clara Valley. The fruit orchards in the Valley were mostly in the time before the invention of the American Teenager. Of course, people had always grown up through their teens, but there was a vast change in the development of young people after work was eliminated from their educational experience. Kids who work grow up understanding the concept of *earning* a living. They make the connection between productive work and the provision of necessities, and are less likely to grow up with the idea that other people will, or should, provide for them. The elimination of work from this group has produced quite different results. There is an old Chinese proverb that says, "Give a man a fish, and you feed him for a day; teach him to fish, and he'll feed himself for a lifetime."

When the orchard farms dominated the Santa Clara Valley, all the kids worked. Early child labor laws excluded agriculture for the common sense reason that all hands are needed at harvest time. It was not until near the end of the fruit growing era in the Valley, in the late 1950s or early 1960s, that labor laws became more restrictive and workers under the age of twelve, unless family members, were prohibited.

When the Second World War broke out, nearly all the young men were enlisted into military service. Most of the remaining work force, including many women who had not previously worked outside of the home, was occupied with war production. That left a huge farm labor shortage.

We have already discussed (in chapter 11) farm labor with regard to the immigration of Chinese and Japanese workers in the middle and late 1800s. Mexican farm laborers did not come to California in significant numbers until about 1900.[171] The immigration of Mexican workers continued until the Great Depression, during which time the last thing the country needed was more people looking for work. However, with World War II, what became known as the Bracero Program began. After Mexico declared war on the Axis powers in June of 1942, the US Congress passed a new law that allowed the contracting of Mexican laborers for agricultural work in the US. Mexican farm labor was considered part of the war effort. Almost all of the Braceros—roughly translated "arms"—worked for the larger growers. For most of the family farms, there was no one left but the kids to bring in the crops. And they did.

On the Pitman farm, the three Pitman children and many neighbor children brought in the crops during the war years. While doing my research for this book, I spoke with a kind woman at the Saratoga Historical Museum and mentioned to her that I had worked on a farm in Saratoga. She asked me which one, and I told her, "The Pitman farm." She replied, "Oh, someone here used to work there." And so I met Ernie Kraule and heard one of the best stories of the era. I'll tell the story here, not only because it paints a picture of the times, but also because it is instructive on several points.

Ernie worked on the farm for two summers, probably in 1943 and 1944 when he was seven and eight years old. His family lived in San Francisco, and Ernie came to Saratoga for the summers. The Kraule's summer home was next to the Harrisons, and the Pitmans knew the Harrisons. During the prune harvest, Mr. Pitman would drive into town in the white truck (yes, *the* white truck) to pick up the Harrison and Conroy kids, and Ernie would come along with them to pick prunes. At the end of the day, Mr. Pitman would take them all back home.

Some people today might call that exploitation of children. To Ernie it was not that at all. He told me, "It was a wonderful way to grow up. Mrs. Pitman was such a gracious woman. She served us lunch, I don't think every day, but quite often. And there was always fresh milk." Ernie is not the only one who told me about the milk and how appreciated it was. This was during the war, and nearly everything was rationed, so milk was not always readily available. "Every week [Mr. Pitman] would set a quota of work that needed to be done, and after the work was done, he'd take us up to Steven's Creek Dam where we all went swimming. They were very good people. They were Christians."

By picking prunes, Ernie earned the money to buy his first bicycle, and all these years later he has remembered the Pitmans with great fondness.

That we all would live in such a way to be so remembered!

Mr. Pitman never turned down a kid who wanted to work. His son, Dave, told me this, and I witnessed it myself. Dave told me the story of a young boy, about six years old, who came with the other kids one year during 'cot season. He was not old enough to be of any help, and the other kids would help him finish his tray of apricots every day, but even he was not turned down. At the end of the week, Dave recalls his dad saying in good humor, "I don't know if I should write you a check or send a bill home for babysitting." He wrote the check. And I have no doubt that the boy's mother and many other mothers greatly appreciated having their kids so constructively occupied during part of the summer.

When the Pitman children were all grown and gone, a Mexican-American family from Hayward came to take care of the harvests. There were three or four adults and several children. The family stayed for the summer during the apricot and prune seasons, then went back home to Hayward. After the new dry-shed was built, they lived in the loft while working on the farm.

Until the end of the agricultural era in the Valley, housing for most of the migrant workers was usually in the form of little cabins at the various farms. On my bicycle ride to the farm, I passed by the cabins that were in the back of the cherry orchard that was around the corner from our house, then rode past the Cox place on Cox Avenue. I would usually see the Mexican families sitting on buckets or squatting around a fire cooking their breakfast in the orchard outside the little cabins there. Farther down Cox Avenue, in an orchard by the railroad tracks, there were more cabins—though I don't remember seeing anyone there.

I was well aware that mine was a far more privileged life than that of the migrant workers' children who stayed in those cabins.[172] I went to work at the farm because I wanted to. Many of them went because they had to. And I'm sure that their stories would be quite different than mine.

The stories I have heard from the children of farm owners also vary greatly. Some, even from relatively wealthy farm families, felt that they were treated as slaves. Others told me it was a good way to grow up. Many were neutral about it—it was just life. Listening to them, it is obvious that it was the character of the people involved more than the economics that made the difference.

In the early days of the family farms in the Santa Clara Valley, a decent living could be earned by doing farm work. Many people worked on a farm for someone else first, learned how to manage a farm, then leased a farm themselves. If the crops were good and the market favorable, sometimes managing a leased farm for a percentage of the gross was a stepping stone toward purchasing a small, usually underdeveloped, farm or piece of land.

As the population in the Valley increased and the land became more and more valuable, the gulf between farmer and farm laborer became greater and greater. Often labor became, not people you knew, but strangers, hired through a contractor to do the work. This was a stark contrast to the way most of the family farms in the Valley had been operated.

At the Pitman farm, after 1959, it was no longer viable for the family from Hayward to make the trip, so Mr. Pitman started hiring local kids again. It was just a few years later that I came on the scene. By that time, more and more orchards had disappeared to make way for housing tracts, and working on a farm had become something of a novelty.

When I was growing up, I did not consider our family to be "poor." We were certainly better off than the migrant workers. We just didn't have money for much other than necessities. There was not money for Little League, and I have to admit I was at least a little jealous of the kids who got to wear those uniforms and play a "real" baseball game. But my dad took us up to the school in the evenings and on weekends and played baseball with us. The other kids in the neighborhood would see us walking up the street to the school with bats and gloves in hand and would come running. Sometimes, we had enough kids to fill five or six positions on both teams. By the early 1960s, it was mostly kids like these, whose parents didn't have the money for Little League or Boy Scouts, and who didn't have another way to get money, that picked prunes and cut apricots. And we were happy to have the opportunity to do it.

In those days, before a little pill hit the market and "family planning" clinics sprouted up everywhere, people had a lot more children than they do now, so there was not a shortage of willing hands for the farm work. Word got around where the work was. I'd tell a few kids, they'd tell other kids, and pretty soon I was flooded with requests. And through it all I learned a very simple and effective business formula.

Next page: The 1943 or 1944 Pitman farm prune picking crew. They are on their way to go swimming at Steven's Creek Reservoir. Dave Pitman is at front center in striped shirt, "Red" Parker is to the right of Dave, Ernie Kraule is behind "Red" and to the right, Carol Norell is in the striped shirt standing in the middle of the truck. That's the *"white" truck in its better days.*

33
The Business Formula

Today, there is no shortage of books written about employment practices, both from the standpoint of the worker and of the employer. After all is said and done, I think the success of that relationship comes down to two very simple matters: desire on the part of the worker, and fairness on the part of the employer. If the desire is not there on the part of the worker, there is dead weight dragging down the operation. If fairness is not there, people feel cheated, negativity spreads, and the operation is undermined. It doesn't matter if you are hiring prune pickers or accountants, engineers or doctors, school teachers or school janitors. If those two things are not present, you've got a tough row to hoe—there are too many weeds, the soil is too hard, and your crop will suffer for it.

I learned this principle quite early. As mentioned earlier, after my first prune season at the farm all but one of the older boys left, so in the remaining years I supplied the prune picking crew. If I hired kids who didn't need the money, or felt they had something better to do, they simply didn't do the work. Furthermore, they were a distraction to the other pickers and caused them to become sloppy. Those who had previously cleaned their row of trees well would begin to leave fruit behind. Those who had never broken the cardinal rule of not throwing fruit would get involved in a fruit fight. All order was lost.

The piecework system eliminates a huge amount of nonsense in the worker-employer relationship. In the mid-sixties, prune pickers were paid 35 cents for picking a box of prunes. The box was either full or it wasn't. If you had fifteen boxes stacked up and they weren't full, some boxes were emptied to fill the others, and maybe you only had thirteen boxes. I know this. It happened to me in my first year of picking. From then on I filled the boxes. If you didn't clean your row, you were sent back to do so. And if you spent your time thinking or complaining about how hard the work was, you didn't end up picking enough prunes to make it worthwhile.[173]

Of course, not all farm work can be paid by the piece. Apricot pickers were paid by the day. Unlike prunes, apricots and fruits that are going to be sold fresh require careful handling or they will be bruised and become worthless. Furthermore, these fruits need to be picked after they have achieved a certain degree of ripeness and discarded if they are too ripe. Getting hired to pick 'cots or any other job that paid by the day was a privilege earned, and at the time I was involved, the kids were well aware of that fact.

During the spring vacation from school after my first summer at the farm, I was hired with a young man named Dave to help with some other chores. One of the first of these chores was to hoe weeds from around the little trees that had recently been planted around the perimeter of most of the property. Dave was five or six years older than I, and he was much stronger. We were set to work early one morning hoeing weeds, and after lunch Mr. Pitman came out to see us. His purpose was to check on our progress and to determine my rate of pay. So the three of us had a discussion. Dave was making $5.00 a day; what would be fair for me? The formula was very simple: for every five trees that Dave hoed, how many did I hoe? Dave and I discussed this for a little while in front of Mr. Pitman, and came to the conclusion that I probably hoed about three and a half trees to his five. So my pay was set at $3.50 per day. It was that simple and I think we all felt good about it.

Everything with Mr. Pitman was like that: straight up and fair. Decades later, when interviewing Dave Pitman for this book, he told me the story of the sale of the Surrey Lane parcel of land in 1954. Mr. Pitman was no longer a young man, and Mrs. Pitman wanted him to cut back on working so hard, so when a man came by and made him an offer for the land, it was considered. After some time, the two came to an agreement. Very shortly thereafter, another man came by and expressed an interest in the land. Mr. Pitman told him that he had just sold it. The man asked, "Are the papers drawn up yet?" "No." He then proceeded to offer him more money. A lot more money. "No." Later, when asked why he didn't negotiate for more money, Mr. Pitman simply replied, "I shook the man's hand." His "yes" was "yes," and his "no" was "no."

But I soon found out that not everyone did business like Mr. Pitman.

By the time my second summer on the farm rolled around, I had a paper route. The young boy who delivered our newspaper was in my sister's class. He was a year older than I and his name was Doug. One day my sister and I were out in the front yard when Doug rode by delivering his papers, and he stopped to talk us. He excitedly told us how he was working hard to get a $300.00 scholarship so that he could go to college at San Jose State. Now to me, that was nearly the equivalent of going to the moon. My folks had never gone to college and I understood that college cost a lot of money. And we didn't have money. But here was Doug talking about earning a scholarship to go to college by diligently delivering his papers every day. If he just did his job right and his customers did not call the newspaper office to complain that they did not get their paper, did not get it on time, or that their paper had been thrown onto the wet grass or in the gutter, etc., he would earn the designation of "honor carrier." If he was an "honor carrier" for three years, he would receive the $300.00 scholarship. Wow! Going to college was possible! Doug was going to do it—and so could I! So I hotly pursued a paper route and within about six months, I had one. I was an "honor carrier" every month for two and a half years. But I wasn't dealing with Mr. Pitman here, and the rules were different. After two and a half years, the newspaper dropped the scholarship offer. They just dropped it. The newsboys were simply told that the paper wasn't offering it any more. There were many disappointed kids. And this was in a day before lawsuits had become as common as house flies in the back porch on a summer day.

Fortunately, by that time I had learned how to earn money and it was not the end of the world. I could find work and put myself through college. So after the farm, I worked at a gas station and from there, like many farm boys in the Valley had done, I went to work in construction. While thus employed, I had the privilege of working with several good men. But I found out something that was not so good. The carpenter's union had a rule that one had to work, through the union, for a minimum of 1,000 hours in a calendar year, or that year would not count toward your pension. It didn't matter that fifty cents an hour was taken out of your paycheck every week for the pension fund. At the end of the year, if you did not have a thousand hours, your slate was wiped clean, all that money taken out of your paychecks evaporated, and you started over. The problem is, building is seasonal, and also goes in boom and bust cycles. So there were relatively few carpenters who had many "good" years. Commonly, a man who had worked as a carpenter for twenty-five years had less than half of those years as qualifying years. A number of men I worked with were getting a little older, and their pension, or lack of it, was becoming a big concern to them.

For a young man just entering adulthood, this was a real eye-opener. But I had an option, and so did the other kids who had worked for Robert Pitman. Though we did not realize it at the time, he had trained us well. While working for him we had picked up many "life-tools" and were not afraid to try anything. So, after getting an education and some further experience, I determined to start my own business.

And I had Mr. Pitman as a clear, unwavering example of how to treat people fairly.

34
The Decommissioning

One afternoon on the farm, I came out of the orchard to the driveway and found Mr. Pitman looking at the back of the yellow truck. He had just arrived in it, and as I got closer I could tell that he was uncomfortable—as if he did not know what to do. This was unusual, for he always knew what he was doing and went about his work with confidence. But I could tell that something was different. So I approached him to see what was the matter. He was fussing with a small chain in back of the truck, and I asked if I could help. He told me that a police officer had pulled him over and had asked him, "Where are the fenders on this truck?" To which (after a pause I'm sure) he had replied, "That's like asking me where my teeth are. They're just not there anymore." Point made, but the problem was not solved, and he was presented with a piece of paper that he would rather not have received. Apparently, the motor vehicle code requires mud flaps, at minimum, if fenders are missing on a vehicle. So he was working on a mud flap—something that would keep the rear wheels from slinging road debris on cars following behind. He was looping the thin, long chain around a protruding rod that at one time had supported the long-missing fender, but he obviously was not satisfied with the results. I did not understand much at the time and remember feeling as helpless as he did concerning the task. No matter how it was done, it just did not seem that our mud flap would be effective. There really wasn't much that a fourteen year old boy could do, and it was not beneficial for me to be there, so after a while of uselessly standing by, I went on to my other chores. I hated to leave Mr. Pitman with his problem, but it was obvious that I could not help.

Not long afterward, a new green truck appeared in the driveway. Well, not a new one—it was actually about fifteen years old—but for the farm it was new, and it looked like a showroom piece parked next to the yellow truck. And it had mud flaps. So the yellow truck was relegated to serve as an orchard truck.

Of course, this was not the first de-commissioning on the farm. Before I came along, both barns—which to borrow a realtor's term, had "matured"—had been replaced after the sell-off of the Surrey

Lane property. As I stated earlier, the property was sold because Mrs. Pitman wanted Mr. Pitman to stop working. I laughed when Dave Pitman told me that, and he did too. No, you were not going to keep the man who said, "You can't have a good day until you sweat" from working. At any rate, the back property was sold and that freed up some cash for the new barns. By the time he was ready to build the barns, the city of Saratoga had incorporated, so a building permit was needed. Mr. Pitman went over to the city offices and asked to get a permit for his new barn. He was told that Saratoga was no longer a farming community and that they would not be issuing permits for new barns. Hmmm… Thankfully there was a chink in the armor of the city's stance. If the old barn came down first, he would be allowed to re-build it. Fair enough. So the man who was supposed to stop working so hard tore down the lower barn, cleaned up all the wood and stacked it neatly, laid out the new building, and put in a foundation the likes of which has always caused me to pity whoever it was that years later had to tear it out. He then framed up the new dry-shed, as he called it, and put in a bathroom. The dry-shed was as sturdy a structure as I've ever seen. It was diagonally skinned with redwood siding, and the floor planks of the loft were of heart redwood that were probably close to sixteen inches wide and three inches thick. Two new concrete apricot kilns with steel doors were also built on the back side, and a concrete loading and staging area was added too.

Up by the house, the upper barn was torn down and a new garage was built that also served as a barn. Other than a large door in the back wall, it was a typical four-car garage with three parking stalls and the last space used as a tool shed. The new garage looked a little out of place. It was very nice—well built with new lumber and serviceable—but the house was built in 1893, the dry-shed looked like Fort Knox, and that new garage had the appearance of the tract-like buildings of the day.

A few years later, there was also the decommissioning of the row of walnut trees that were all along the front of the property. These gave way to the widening of the highway shortly after I arrived at the farm. The walnut trees were never a big money maker—I read in an old journal Mr. Pitman kept that in 1946 he sold the walnut crop from those trees for $172.25. That wasn't much considering the effort that was expended on them, but every bit helped.

The decommissioning of the walnut trees brought with it the reminder that there were always people who would try to take advantage of the farmers. At the time of the road widening, an attorney from the governing agency responsible for procuring rights-of-way showed up at the farm. He told Mr. Pitman that the road was going to be widened and since it would benefit him so very much, would he not sign over the deed for the needed frontage? Although I was not there to see that exchange, I can visualize how Mr. Pitman would have handled it. I'm sure he paused first, and I'm told his words were something close to, "I'm not in the habit of giving quarter acres away, but you can buy it if you want it." They bought it.

Actually, the farm itself was slowly being de-commissioned. Property taxes, in those days before California's Proposition 13, were taking a bigger and bigger bite every year.[174] A farmer could only hold on so long, and almost all the other farmers had thrown in the towel long before.

Being older now, I realize that it was quite likely that all these things were going through the old farmer's mind as he was draping that chain in back of the wheels of the yellow truck.

Some things aren't so easy.

"The Yellow Truck" was very similar to this one

35
When the Cat's Away…

During the prune season, Mr. Pitman would take a load of prunes to the dehydrator nearly every afternoon after lunch. The Sunsweet dehydrator was on Railway Avenue in Campbell, and it took him about an hour and a half to make the trip. With two old orchard trucks available to a bunch of 13 to 15 year olds, it is not difficult to figure out what happened when he left.

Let the Games begin! The capitalization of "Games" is not a typo. No, these were serious contests of chivalry, and they tested the skills of the "young bucks." Some of the other boys learned how to drive the trucks, and we lined them up on rows that had already been picked and had drag races. It was of no consequence that the outcome of the race was a foregone conclusion. The difficulty of shifting the white truck made it impossible to get into second gear very quickly, so the yellow truck always got to the end of the row first. But the roar of the engines (actually their backfiring) was exhilarating and it was great sport.

After the orchard was picked to a certain point farther away from the house, there were some rows with a few trees missing, so we also set up a slalom course. Great fun! Autocross on the cheap!

Then there was orchard surfing. This was when some brave, or looked at another way, stupid, boy would stand up in the back of the truck and see if he could keep his balance while being driven around the orchard. In time, orchard surfing was mastered, so we had to make it more challenging. The goal eventually became one of trying to knock the surfer off. The driver would dodge in and out of the trees picking his spot for a sharp turn or searching for a tree branch that was hanging at just the right height. It seems not to have entered into our thought process, or if it did, it didn't stay, that someone could end up getting run over or otherwise hurt. It was just too much fun and we used our best (?) judgment as to what we could handle and what we could not.

Another event, a take off of the slalom races, was orchard skiing. This was a favorite because it was not considered nearly as dangerous as the surfing combat had become. Orchard skiing could only be done with the white truck because it was the one that had a tailgate. The skier would grab hold of the tailgate with both hands, lean back with arms fully extended, put his feet forward, dig his boot heels into the soft dirt, and lock his legs straight. Another boy would then drive the truck through the slalom course. It was water skiing without the water and without the skis. The boys waiting for their turn cheered on the skier. Of course, orchard skiing also became competitive, the goal being to see who could hold on the longest while the course was run. Some of the turns were just too high speed—maybe 15 mph—and the skier was forced to go too wide. He inevitably lost his grip and took quite a tumble, which was to the great delight of the fans. Then the next boy took his turn, and we all laughed at him when he "lost it." We tried very hard to master this one, but to no avail. A good driver could always dump his skier on certain turns. Thus we had our first lesson in centrifugal force, and later, for those who took physics, that part of the curriculum was a snap.

At the appointed time, the Games were stopped. The boys then went around the orchard, and with their feet did their best to cover over the worst of the tire marks, the likes of which would never have been made by the calm driving around in the orchard trucks while gathering the boxes of prunes. Everything was always back in order by the time Mr. Pitman drove up the driveway, and all the boys were under their respective trees enjoying their picking a little more than usual.

Later, looking back on it, I realized that there was absolutely no way that Mr. Pitman did not know what we were doing. Mrs. Pitman may have been hard of hearing, but she wasn't deaf, and I think she had to hear at least some of the commotion. Besides that, not all the skid marks in the dirt could be perfectly disguised.

But then there was the trailer incident…

36
The Trailer Incident

There is fundamental problem that arises when young boys have too much time on their hands. Boredom or restlessness can set in, and then… well, things happen. The trailer incident is a case in point. The trailer incident could have been put with the previous chapter, but is deserving of its own for reasons which will become evident.

Since the prune season was in the midst of the hot weather, and by noon the pickers needed a good break, we went go down to the dry-shed, where it was always cool, to eat lunch. Mr. Pitman always took exactly an hour for lunch, and on the hot days we did too. But after about a half hour in the dry-shed, we had eaten lunch and were cooled off and revitalized. Not wanting to go back to work quite yet, but needing something to do, some of the boys started to look around. The farm machinery and implements were not of much interest to most of the boys, and after they had explored the back acreage, some got bored and mischievousness set in. A couple of the boys who were soon to start football practice wanted to get some early practice in. How better to do this than to practice tackling sheep? I strictly forbid the proposition. If one of the sheep got hurt, I explained, it would not be good. So there was a little trouble brewing, but no outlet.

One day when we entered the dry-shed for lunch, a new arrival was there. It was an old open trailer, the kind you tow behind a car. It was about ten feet long, with two wheels, and had a thin steel frame with wooden planks for the floor and sides. The sides extended up about two feet, and the tires were good and well-inflated. It was newer and in better shape than the trucks and most of the other equipment on the farm, and I suppose it was for that reason that we found it quite attractive. Several of the boys checked it out real well, and then we all sat down to eat lunch.

After some time, one of the boys got up and started playing around with the trailer, lifting up the tongue and rolling the trailer back and forth a little. Another boy had gotten up and was looking out the large door of the dry-shed toward the apricot hill. It was a beautiful day, and we were all by now refreshed and almost everyone was starting to get up and move around a little. Some boys began wondering aloud how we could take a ride in the trailer. The boy looking out on the apricot hill calmly called another over to have a look. A suggestion was put forth, which at first was dismissed outright, but in time the talk became serious. Other boys went over to the door and looked out at the apricot hill, then looked back at the trailer, then looked out at the apricot hill again, and then back at the trailer. The air became electrified. The next thing you knew there were six or seven boys pushing the trailer up the hill.

This had not been thought out very well. Granted, the apricot orchard had been abandoned some years previous and most of the apricot trees on the hill were gone, but there were still several left, and the dry-shed was not far from the base of the hill. By the time we got the trailer to the top of the hill, which was no easy task, we were pretty tuckered out. The steepness of slope required the full effort of two boys just to hold the trailer in place, so decisions were being made rather quickly. The fact that we could not steer the trailer, although a consideration, seemed to be of lesser importance than the ride that awaited.

Coordination was needed. We realized that everyone had to get in at once, because the trailer would begin its descent as soon as the two holders let go. So we devised a boarding plan like that used by a team of bobsledders. There was a brief last-minute discussion as to the dangers involved; but we were already at the top of the hill, there was only one direction to go, and we were all tiring from holding the trailer in place. So in jumped most of the boys. Immediately a problem we had not anticipated became evident—the boys holding the trailer could not bear up against the additional weight in the trailer. The trailer was on the move. One or two boys never made it in. Whether that was due to superior intelligence or logistics, I don't know. As the trailer began its decent, the gravity of the situation became apparent and boys began bailing out. When they did, another unanticipated problem was encountered, that is, when someone bailed out, he pushed off with his feet against the side of the trailer, and since every action has an equal and opposing reaction, the trailer was sent in a different direction. With speed gaining (and it all happened very fast), the trailer was rapidly and radically changing course as it went down the hill. I stayed in the longest, hoping that I could somehow steer by leaning, but even the chief idiot realized there was no saving this one and bailed out.

The trailer was then flying down the hill on its own and was headed straight toward the only apricot tree at the base of the hill, which it struck squarely. The tree stood fast, but the trailer did not fare as well. The impact sent splintered boards flying far and wide in fan-like fashion. Hollywood explosive engineers could not have done a finer job. I mean, the thing just about disintegrated. It was absolutely the grandest spectacle any of us had ever seen, and was only matched by all of us rolling around on the ground holding our bellies from laughing so hard. After a while, some of the boys started to get up to dust themselves off, but when we looked at one another, everyone was on the ground again trying to keep their entrails intact.

Eventually, some of the seriousness of what we had done started to set in, but at the same time it was so hilarious that the fits of laughter just could not be stopped. As the lunch hour drew to a close, we knew that we would have to face Mr. Pitman. Though we tried our best to begin to compose ourselves, we found it impossible to do so. All that needed to happen was for two boys to exchange glances, and the hilarity broke out all over again. I tried to calm everyone down, but in fact was no better off than any of the others. The whole thing was just too funny. We gathered up what remained of the trailer, put it in the dry-shed, and with probably the greatest self control ever exercised by any of us, began our ascent up the road to the farmhouse to tell Mr. Pitman. Boys still broke out in laughter now and then but were harshly rebuked by the other boys, who then burst out in laughter themselves. It was not a good situation, and our prospects did not appear any brighter than our intelligence.

I knew that I was responsible for the entire matter and was determined to bear the responsibility for it, but we were all going to face Mr. Pitman. We met him out on the driveway shortly after he came out of the house. How we all kept our composure as I confessed to him what we had done, I don't know. I suppose we saw things in a different light once we were in his presence.

After hearing me out, Mr. Pitman looked at all of us and said, "Well, I just got that trailer yesterday, and I didn't have any idea what I was going to do with it." I apologized again, and the other boys apologized too. There was no more laughter, and we all went back to work.

We never saw the trailer again. The next time we went down to the dry-shed for lunch, it was not there. I don't know what Mr. Pitman did with it, but no part of it was anywhere to be found. And he never said another word about it.

Blessed is he whose transgression is forgiven, whose sin is covered.
—King David

37
Eat All You Want

The variety of fruit trees Mr. Pitman had planted over the years was really something special, and he was happy to share it all with anyone who came to the farm. As mentioned earlier, the original prune orchard was planted in the 1880s, and though the trees continued to produce extremely well, they very much looked their age. They had sent their roots deep into the loamy soil, but the trunks of many of the trees were termite eaten and rotting away to such a degree that one wondered how some of them remained standing. Of course, some didn't, and from time to time Mr. Pitman replaced the trees that fell over with fruit trees of other varieties. By the time I came to the farm, there were Bartlett Pears, Babcock Peaches, and Santa Rosa, Black, and Elephant Heart Plums in the orchard. Near the house, there were Concord and Thompson Seedless grapes, a Loquat tree, an apple tree, and in the garden, blackberries, tomatoes, and cantaloupe.

Words cannot express the goodness and refreshment of these fruits to one who had been working hard and was in need of nourishment. Also, picked and eaten ripe and fresh, fruit tastes, and in fact is, very different than that which is on a grocery store shelf. The difference in vitality between the two cannot be compared. Add into the equation the ability of the body which has been greatly exercised and has sweat profusely to absorb such nutrition, and you have a most wonderful phenomenon. Those who have never known this have missed out on a great gift.

Dave Pitman tells the story of how his mother would sometimes become very upset with his dad when she had worked hard to prepare a nice dinner and he was not hungry because he had been eating fruit all day. As for me, many times I did not take a lunch to work, and at other times it was quite minimal. I would simply go pick what I wanted and take it down to the dry-shed to eat. Most of the kids that I picked prunes with came from families who either could not afford, or had not given in to, the then-new refined foods that train our taste buds to despise healthier foods, so they also enjoyed the fresh fruit. We all ate as much as we needed, and in that way were well energized to do our work.

In the last summer that I was there, which turned out to be the last year of the farm's operation, I invited a certain friend to come pick prunes with us. This boy had no financial need and I knew that that was not good, but he was a close friend and he wanted to come, so I allowed him. He drove up in his white sports car late one morning mid-season and I showed him around, gave him his buckets, and assigned him a row. After lunch time, as we were getting ready to go back to work, with very wide eyes he asked me if it was really true that we could eat all the prunes we wanted. Some of the other boys looked at me with, "doesn't he know?" written all over their faces. I tried to calm his excitement a little by advising him that it would be good to use moderation, and some of the other boys chimed in with a word of warning, but he grew even more excited, so we didn't press the point. Rather, we made sure he knew where the bathroom was.

My dear friend made many trips down to the dry-shed that afternoon. And I don't think he came back the next day.

38
Free Pipe

The summer after I began working at the farm, road construction for the widening of Saratoga-Sunnyvale Road in front of the farm began. On the opposite side of the road, just up from Miljevich Drive, there was a long guardrail that needed to be taken down. The guardrail was made of two inch steel pipe painted white and was typical of the style of rail that was installed in the 1920s and 30s beside the creeks and over the small bridges in many places in the area. Some of it still survives on Quito Road, and I'm sure in some other places too.

In those days before plastic PVC pipe, Mr. Pitman saw an opportunity for irrigation pipe and obtained permission from the road-work foreman to salvage the pipe. So he took me out with him to harvest that crop. With traffic whizzing by on the highway, we used a hand pipe cutter to cut the horizontal lengths of pipe that were between the vertical posts. The lengths we were able to get were about seven feet long, and we got a good stack of them. At the time, I was only twelve or thirteen, and was not much help cutting the pipe, although I remember putting my hand to the task. So he cut the pipe and I loaded the truck, and by the end of the day we had a full load of pipe, which we drove down to the dry-shed.

I've previously mentioned the young man, Dave Fowler, who worked with Mr. Pitman and me that year. Since I did not have sufficient strength, it fell to Dave to put threads on the ends of all those pieces of pipe. In Mr. Pitman's words, threading pipe was, "The kind of work that made an old man out of a young boy." I'll never forget coming into the dry-shed from my job of watering the trees and seeing Dave set up at the work bench by the big door sweating as much as a man can sweat as he threaded the pipe by hand. He threaded the entire stack.[175]

The fencing for the pasture had been procured in similar fashion. In 1936, the Golden Gate Bridge had been completed, and in the following year, the Oakland Bay Bridge. These were huge accomplishments and to commemorate them, San Francisco was chosen as the site for the 1939 World's Fair. The Fair was held on

Treasure Island, which originally was just a rock outcropping similar to Alcatraz Island, but Treasure Island was expanded greatly by dredgings and fill from the bridge projects. The new, flat land provided the fairgrounds. After the Fair, the island was converted into a military base, and almost all of the Fair-related infrastructure had to go. Much of it was available to anyone who would cart it off, and that's where Mr. Pitman came in. So, the pasture got first-rate fencing and Mr. Pitman himself got several green denim uniforms that had been worn by the Coca Cola Company employees during the fair. I'm told that he wore green for years.

No, the old farmer was not the kind to spend money unnecessarily. Food he had. Time he had. Cash… that was another matter. In later years, when he decided to build a dam over the creek in the "L" parcel, he sent his boys with the truck to the creek below the gravel quarry to get aggregate for the concrete. Of course, the concrete was all mixed by hand. When he needed a gin pole, he just went up into the hills and found a fallen tree that looked about right.

Things were different then.

39
Solitude

A fair amount of farm work is one on one. By that I mean you are alone with the task to be done, and it either gets done or it doesn't. And it's either done well or it's not. And many times only you know—at least right then.

Cutting 'cots was social, and picking prunes was somewhat so—you at least saw the other pickers. But most of the work on the farm was not like that. You were on your own. Much of the work was not mentally taxing, so frequently there was available "mind-time." I remember watering the trees out on the back hills where the water ran slowly, and watching the sub-chasers fly overhead. These were the Navy P3C airplanes, and they were headed into Moffet Field Naval Air Station after a day of searching for Russian submarines. The P3Cs had taken the place of the blimps by the early sixties, and there was a constant changing of the guard as they patrolled the coastal waters. It was the strangest thing being out there in such a beautiful place, looking at the blue sky and the oaks on the nearby hills, with the afternoon breeze wafting the pleasant, complex aromas of earth and vegetation, and realizing that we were so close to having none of it. Tension was very high in the early and mid-sixties between the US and Russia. We had practiced "duck and cover" in grade school (as if it would do any good in a nuclear blast) and the new neighborhoods all had yellow poles in them, which were topped with sirens in case of an attack. Seeing the P3Cs sometimes made the task at hand seem insignificant. But then, in the harmony of that environment, the P3Cs seemed insignificant too.

There is a lot of noise that we are all subjected to every day. There is the physical noise—the radio or TV, telephone, traffic noise, billboards, unnecessary and foolish talk, etc. And there is mental noise, much of which is leftover from the onslaught of physical noise we hear. Sometimes it is good to get away from these influences, and the time on the farm was that for me and for several other young kids. While going about our work, we learned to hear and see things that are often overlooked. At the same time, there was always more to do on the farm than there was time to do it, and in such an environment one learns to hate idleness. So we learned both the value of quiet time and the value of using our time wisely.

Some of the farm kids expressed their appreciation for this training, and many found constructive ways to spend their time when the mind did not have to be fully occupied with the tasks they were doing. One of the boys worked out math problems in his head while occupied alone with more mundane tasks. I learned to whistle. Whistling did not come naturally to me and took a substantial amount of practice. For many days I could get no sound but the blowing of air. But I was determined, and one day it just came out.

On the farm, many of us also subconsciously came to the conclusion that it was easier to do a good job than it was to find excuses not to. Besides, there is a sense of satisfaction that comes from looking back at an orchard after placing all the props the best you can under the heavy-laden branches, knowing that those trees had the best chance that you could give them of surviving the birth pangs of the coming harvest.

In the end, we are all alone in many respects. It is easy to think that no one else sees, and that it does not matter what we do or think.

But it is not so.

40
“But Mom, It’s My Home!”

Some farm work is boring, some is rewarding, and some is just plain hard. The aching muscles kind of work—heavy, *hard* work. Although Mr. Pitman was good at assigning tasks that you could handle, sometimes there were tasks that just needed to be done, and whether you were up to the job or not was immaterial—you just did it.

It is one thing to be tired out. That’s good. But to strain yourself is not good. Once, probably when I was about fourteen, I developed a sore back and was just a bit physically broken down. I had come home too sore and my mother caught eye of it. I told her I was OK, but I wasn’t, and as mothers do, she knew it. This continued for a few days, and at one point I made a comment about not feeling well. She immediately responded with something like, “That’s it, no more. You’re not going back there any more. It’s too much.” I tried to reason with her and told her it was no big deal, that I’d be fine and so forth, but she wasn’t buying and stood her ground. I pleaded a little more, but she was set. Eventually I broke into tears and blurted out, “But Mom, it’s my home!”

I’m sure that shocked her, but after the initial shock I don’t think she was offended. She knew I loved and respected her and my dad, so that wasn’t the issue. She looked at me and paused for a long time, and then nodded her head and said, “Okay.” We agreed that I would rest for a few days, and in the future not try to do things that were too much for me.

From then on I was more careful.

I mentioned in a previous chapter that Dave Fowler threaded a stack of pipe by hand and how hard that work was. I didn’t say in that chapter that at the time he had another job. It was at a bank, doing paperwork. But there he was on the farm doing some of the hardest work a man can do. Looking back on it, I can only guess that he loved being on the farm and with Mr. Pitman as much as I did.

We were not the only ones.

If you have ever known the scent of freshly turned dirt, run your hand through a newly sheared lamb's fleece, seen a newborn lamb skipping, heard the symphony of a pasture in midsummer, had hours to watch the hawks soar while going about your work, smelled the fragrance of a blossoming fruit orchard, walked through an orchard early on a damp morning in the spring when the mustard is in full bloom, tasted fresh fruit or sweet well water on a hot summer day when tired out from hard work, looked out on an orchard after you've pruned all the trees and cleaned up all the cuttings, or enjoyed the aroma of fresh cut wood, then you'll know why we stayed.

If all that wasn't enough, the privilege of being with Mr. Pitman easily made up the difference.

41
Transitions

As sure as wave comes after wave on the Pacific shore, so change follows change. The Ohlone gave way to the Spanish, the Spanish gave way to a new Latin American nation, and the Mexican Californios gave way to the Americans. Likewise, cattle ranching gave way to wheat farming, wheat farming gave way to orchards, and the orchards gave way to more modern technology. In all cases, there was simply another wave that came in and covered over the remains of the preceding wave. And so it continues.

The Santa Clara Valley was too well situated to not become a major point of commerce on a growing West Coast. Economists had written for decades before Hubert Bancroft published his great work, *History of California* (1890), that industry would soon surpass both mining and agriculture in the state. This was inevitable, but the death keel to the orchard industry in the Santa Clara Valley was the state property tax structure, which allowed real property to be reassessed each year. Many people were unable to pay those taxes and lost their homes or farms. Eventually, a taxpayer revolt resulted in the passing of State Proposition 13, which prohibited reassessment of real property unless ownership changed, but that was in 1978, and the farms were long gone by then. By the middle to late 1950s, property taxes in the Valley were higher than the value of the crops that a farmer could produce on his land. Then they doubled; and doubled again.

The farmers moved on or retired. Many farmers could not entirely walk away from the soil that they themselves were rooted in, so they sold off all but a small parcel which was usually an acre or less. Some of the apricot growers moved to Patterson, the new apricot growing region, which is in the Central Valley just the other side of the Santa Clara Valley's eastern mountains. Likewise, many of the Valley prune growers moved to the Yuba City area, north of Sacramento, which became the new prune growing region. Seagraves, Defiore, the Miljevich brothers, Ljepava, Lipscomb, Russell, and countless others sold out.[176] In 1968, the Pitman farm was one of the last to go.

Before Mr. Pitman left the place, there was something I wanted to give to my old friend. During my time on his farm, I had developed an affection for Model A Fords because I had learned so much from driving and tinkering with his old Model AA orchard trucks. A friend of mine had restored a Model A of the same vintage, and lent it to me one afternoon. It was a special treat to drive up to the farm with my girlfriend in a shiny, well tuned Model A sedan and offer Mr. and Mrs. Pitman a ride. My two old friends sat in the back seat and giggled like a couple of school kids the whole way. Afterward, Mrs. Pitman took my future wife inside the house and taught her how to make an apricot pie, while Mr. Pitman and I talked about I don't know what. I just remember him looking at me, and with a twinkle in his eye saying, "Tim's sparkin'."

That was the last I saw of Mr. Pitman at the farm. I went to see him later when he was living in a rental house while he was waiting for his new home to be built on a small piece of the farm property. He seemed to me like a fish out of water. He was no more fit for living in a housing tract than he had been for a desk job decades earlier.

After the Pitmans moved into their new brick home on Rice Court, I'm sure their neighbors got more than a few chuckles out of the old farmer. He built a low chain link fence around the front yard which at first glance looked normal enough. But upon closer inspection you could see that the top rail of the fence had valves in it. The top rail was in fact the irrigation pipe for watering the yard. The vegetable garden and Boysenberries were also in the front yard—not your typical landscaping for what had become an upscale neighborhood. And I am sure there were other things he did that revealed his roots.

The last time I saw him was after we had moved out of the area and I brought my two year old son, James, with me to go visit him. I was greeted at the front door by Mrs. Pitman who was delighted to see me, and called out in her high voice, "Bob!"

Mr. Pitman was not in good shape by that time. He walked with a slow shuffle and trembled with Parkinson's, but his greeting was as warm as it had ever been.

We had quite a talk—the heart to heart kind—and enjoyed each others' company until my dear little James, who had been very patient, became not so. So we bid each other farewell.

It was a very slow and quiet drive home.

My, how I loved that old man!

A Memorial to Robert Freeman Pitman

He never spoke to me about the things of God
As he had to his children and some others,
Yet he was the most reverent man I knew.

Instead, he showed me how to use a shovel,
Shake prunes from trees, drive the truck,
And do a hundred other things around the farm.
It was in these humble tasks that I knew a godly man.

From his mouth I never heard a boastful word,
Rather at times, and with a chuckle, “Stupid farmer.”

He lived as one who knew that he would give an account,
And but once did I see him lose his peace.

A lot of time has gone by since my years with him,
And though my own life has fallen short
Of the example that was given to me,
It is still my grand ambition
That I would be like him.

Epilogue

From the time I left the farm, I thought there was a worthy book that could be written about that time. That was forty-three years ago. One day, however, while driving for an extended time, this book began to unfold in my mind, and a couple of days later I wrote down the chapter titles pretty much as they are. After outlining the chapters, I was convinced that the story was worth telling.

I did not desire to merely write of old times and ways, but rather wanted to try to put down something of what I had learned and experienced on the Pitman farm that was of lasting value.

In the spring of 2009, I made a trip to visit my mother who still lives in the area, and decided to try to look up Dave Pitman, Robert Pitman's younger son, while I was there. I had seen him on a few occasions while on the farm, but that was the extent of our relationship at that time. I called Dave, told him what I was doing, and asked him if I could interview him concerning his dad. He was more than happy to help in whatever way he could, so we made an appointment for the next day.

After speaking with Dave, and later with some other people who knew his dad, any remaining doubt about whether or not I should go ahead with the book evaporated.

There was a lot about Robert Freeman Pitman that I did not know. He did not like to talk about himself, nor did he dwell on the past. He spoke to me mostly about how to do things; but for the greater part I just observed him. He did tell me a little bit about his children and that he had been a plumber in San Francisco. Once, when I told him that I wanted to take some time off to go on a backpacking trip in the Sierras, his face lit up and he told me something of the pack-horse trips he had taken into Yosemite as a young boy with his father. We talked about it a little but did not need to say much. We understood each other very well.

But in fact, I knew very little about him other than the example that he had shown me. That was the important part, and that life-example was underscored by all those who knew him that I later had the privilege of speaking with.

When I first sat down with Mr. Pitman's son, Dave, more than four decades after leaving the farm, and heard that his dad had gone to the University of California (there was only one at the time, in Berkeley), it came as quite a shock. The man who had on occasion referred to himself as a "stupid farmer" was actually very well educated. He came from a family of educated people—his father was a graduate of USC, his brother, Paul, was the men's dean at San Jose State College during the 1940s, and his other brother, John, was a school superintendent in Turlock, California. The farmer himself helped his kids with their calculus homework.

Mr. Pitman had intended to become a teacher of science and mathematics, but in his senior year at Cal he decided that that was not for him. What he really enjoyed was working with his hands. So he dropped out of school, and after about fifteen years working as a plumber in San Francisco, he became a farmer in Saratoga.

Though he lived a most humble life, Robert Pitman was also somewhat a man of means. The Pitmans apparently had purchased the farm and the Surrey Lane annex outright, and later, when a neighbor passed away having no children, she left the Pitmans her farm.

But you never would have known these things. He was simply Bob Pitman, the farmer.

Though he never taught science and math, Mr. Pitman taught many people how to live. Mostly by example. When the influx of GIs arrived after the war and began families, many of the parents in the area would drop their children off for Sunday school at a local church. Bob Pitman was there, saw a need, and filled it. Placed in front of him were many young adults who were themselves in need of shepherding. They had left family back east, had no roots here, and needed not only a father figure, but also that same inner compass that he had. He would go out to the car to talk with them and invite them to an adult Sunday school class, which began when the first of these parents agreed to come. He spoke to them about the people in the Old Testament stories of the Bible, and during those rapidly changing times when the family unit was beginning to break apart all across the country, many were helped. He provided stability, and that anchor of the soul that he had, he shared with them.

After that first interview with Dave, my mind was reeling. There was quite a bit more to Robert Pitman than I had known, and it was difficult for me to process all the new information. But the next morning I woke up with the following:

"(He) made himself of no reputation, taking the form of a bond servant." Although the quote speaks of Christ, it was this same Spirit that was at work in Robert Pitman.

Appendix I: Modern Prune Harvesting

First photo: the first truck drives up to a tree and the arms under the truck grab the tree trunk. Second photo: the second truck drives up and connects with the first. The tree is shaken. The rate is about three trees per minute. Last photo: the prunes in the shaker are being unloaded into boxes. The yield from these trees was about 250 to 300 pounds per tree.

The number of workers needed to harvest prunes today is a small fraction of what was needed when they were picked by hand.

Trees in a modern orchard are kept much smaller than in times past because smaller trees are easier to work. For other fruits that are picked by hand from the branches, the smaller trees are also much safer. Gone are the days of pickers standing on the tops of rickety twelve foot, three-legged ladders.

LA1002 KUBOTA
LINDAUER
LINDAUER

Appendix II: S. Clara Co. Land Grant Chart

	Name of Rancho	Location	Granted to	Granted by
A	Agua Caliente	S of Mission San Jose	Fulgencio Higuera	Alvarado
B	Ausaymas	Pacheco Pass area	Francisco P. Pacheco	Figueroa
C	San Felipe	Pacheco Pass area	Francisco P. Pacheco	Gutierrez
D	Bennett Tract	Santa Clara	Narcisco Bennett	Pico
E	Canada De Pala	in hills SE of SJ	Jose Jesus Bernal	Alvarado
F	Canada De Felipe y Las Animas	N of Morgan Hill	T. Bowen, N. Daly	Casarin
G	El Corte de Madera	Los Altos	D. Peralta, M. Martinez	Figueroa
H	El Potrero de S. Clara	S. of SJ Airport	James A. Forbes	Micheltorena
I	Embarcadero de S. Clara	Alviso	Barcillia Bernal	Pico
J	Enright Tract	Santa Clara	Francisco Garcia	Micheltorena
K	Juristac	N. of S. Juan Batista	A. & F. German	Castro
L	Laguna Seca	Morgan Hill	Juan Alvirez	Figueroa
M	La Polka	E. of San Martin	Ysabel Ortega	Figueroa
N	La Purisima Conception	Los Altos Hills	J. Ramon, J. Gorgonio	Alvarado
O	Las Animas	South Gilroy	Mariano Castro	Marquina
P	Las Uvas	NW of Morgan Hill	Lorenzo Pineda	Alvarado
Q	Llano del Tequesquite	S tip of SC County	Jose M. Sanchez	Castro
R	Los Capitancillos	New Almaden	Justo Larios	Alvarado
S	ditto	New Almaden	Justo Larios	Alvarado
T	Los Coches	San Jose, near SJCC	Roberto Balermino	Micheltorena
U	Los Huecos	E. San Jose Hills	L. Arenas, J. Roland	Pico
V	Los Tularcitos	N. Milpitas	Jose Higuera	Sola
W	Milpitas	Central Milpitas	Nicolas Berryesa	Chabolla
X	ditto	ditto	Jose M. Alviso	Castro
Y	Ojo de Agua de La Coche	Morgan Hill	Juan M. Hernandez	Figueroa
Z	Pala	E. San Jose	Jose Higuera	Castro
AA	Pastoria de Las Borregas	SE of Moffett Field	Francisco Estrada	Alvarado
BB	Posolmi	Mountain View	Inigo (Ynigo)	Micheltorena
CC	Quito	Saratoga to Sunnyvale	J. Noriega, J. Fernandez	Alvarado
DD	Rincon de Los Esteros	N of San Jose Airport	(Juan?) Ignacio Alviso	Alvarado
EE	ditto	West Milpitas	ditto	ditto
FF	ditto	Alviso/Milpitas	ditto	ditto
GG	Rincon de San Francisquito	Mt. View/ Palo Alto	Jose Pena	Alvarado
HH	Rinconada de Los Gatos	Los Gatos	S. Peralta, J. Hernandez	Alvarado
II	Rinconada del Arroyo de SF	E. Palo Alto	Maria Antonia Mesa	Alvarado
JJ	Salsipuedes	Watsonville	Manuel Casarin	Figueroa
KK	San Antonio (Mesa/ Dana)	S. Los Altos	Juan P. Mesa	Alvarado
LL	ditto	ditto	ditto	ditto
MM	San Francisco de las Llagas	San Martin	Carlos Castro	Figueroa
NN	San Francisquito	Stanford Univ.	Antonio Buelna	Alvarado
OO	San Juan Bautista Narvaez	S. of Willow Glen	Jose A. Narvaez	Micheltorena
PP	San Luis Gonzaga	Pacheco Pass	J. P. Pacheco, J. Mejia	Micheltorena
QQ	San Vicente	New Almaden	Jose R. Berryesa	Alvarado
RR	San Ysidro	E. of Gilroy	Ignacio Ortega	Arrillaga
SS	ditto	ditto	ditto	ditto
TT	Santa Teresa	Santa Teresa area	(Jose) Joaquin Bernal	Figueroa
UU	Solis	Mt. Madonna/ Gilroy	Mariano Castro	Figueroa
VV	Ulistac	S. of Alviso	Marcello Pio, Cristobal	Pico
WW	Yerba Buena	Evergreen area	Antonio Chabolla	Figueroa

	date of Grant	date of US Patent	Patent to	Acres
A	1839	1858	Higuera	9,534
B	1833	1859	Pacheco	
C	1836	1859	Pacheco	35,504
D	1845	1871	Mary Bennett	358
E	1839	1863	Bernal heirs	15,807
F	1839	1866	C. Weber	8,465
G	1833	1858	M. Martinez	10,069
H	1844	1861	R. Stockton	1,939
I	1845	1936	Bernal heirs	196
J	1845	1866	J. Enright, F. Garcia	710
K	1835	1871	R. Carlisle, Sargent	4,533
L	1834	1865	Bull, Fisher, Murphy	17,426
M	1833	1860	B. & M. Murphy	4,166
N	1840	1871	Juana Briones	4,442
O	1802	1873	J. Sanchez heirs	26,842
P	1842	1860	B. & M. Murphy	10,445
Q	1835	1871	Vicente Sanchez	16,016
R	1842	1871	Guadalupe Mining Co	1,111
S	1842	1865	Charles Fossatt	3,360
T	1844	1857	Naglee, Sainsevrain, Sunol	2,682
U	1846	1876	J. Hornsby, J. Roland	39,229
V	1821	1870	Higuera family	4,377
W	1834			
X	1835	1871	Alviso heirs	4,457
Y	1835	1860	Martin J.C. Murphy	8,833
Z	1835	1866	Charles White family	4,425
AA	1842	1865	Martin Murphy Jr.	9,066
BB	1844	1881	Campbell, Wackinshaw, Ynigo	1,695
CC	1841	1866	M. Alviso, Fernandez family	13,152
DD	1838	1862	Ellen E. White	2,308
EE	ditto	1872	Alviso family	2,200
FF	ditto	1873	D. Alviso, Berryesa family	1,844
GG	1841	1868	S. & T. Robles	8,518
HH	1824	1860	S. Peralta, J. Hernandez	6,609
II	1841	1872	Maria Antonia Mesa	2,229
JJ	1834	1861	James Blair and several associates	31,201
KK	1839	1857	H & W Dana, J. Weeks	3,541
LL	ditto	1866	Mesa family	885
MM	1834	1868	B. & D. Murphy	21,659
NN	1839	1868	M. De Rodroguez	1,467
OO	1844	1865	Jose A. Narvaez	8,979
PP	1843	1871	Juan Pacheco	48,821
QQ	1842	1868	Berryesa family	4,438
RR	1810	1867	John Gilroy, J. Martin	4,460
SS	ditto	1868	Quintin Ortega	4,438
TT	1834	1867	Agustin Bernal	9,634
UU	1833	1859	Castro heirs	8,875
WW	1845	1868	J. Hoppe heirs	2,217
XX	1833	1859	Antonio Chabolla	24,342

Appendix II: Land Grant Chart Notes

Note	Notes
A	"Hot Water." Grandson of I. Higuera, an Anza Expedition member. Mostly in Alameda Co.
B	Ausaymas is an Indian name. Pacheco was a Spanish Soldier; Treasurer at Monterey
C	Combined with Asaymas
D	Mary Bennett was an American who came overland from Georgia in 1843
E	A very large parcel went to his attorney, Frederick Hall. Hall wrote History of San Jose, 1871
F	Weber Ranch
G	Translated means, "The Lumber Cutting"
H	Forbes was British Vice-Consul to Alta CA; sold to Commodore Stockton in 1847
I	Includes original docking area on the Guadalupe River for the Santa Clara Valley cow hide trade
J	Enright was an American who came overland to CA in 1846
K	
L	Translated means, "Dry Lake." Sold in 1845 to Fisher, an American merchant from Boston
M	Re-grant of a portion of the Spanish grant called Rancho Ysidro given to her father
N	J. Ramon, J. Gorgonio were Mission S. Clara Indians; sold to Briones in 1844 or 50; note in ch 3
O	Spanish grant
P	"Uvas" means "Grapes." Named after the abundance of wild grapes in the area
Q	Llano del Tequesquite means "Alkali Plain"
R	Part of New Almaden mines; some of the most hotly disputed claims in US Supreme Court history
S	
T	R.B. was a Mission Indian. Lived on land since 1836. Sold to Sunol in 1847 who subdivided it
U	Translated, "The Hollows." So named due to mountainous topography
V	Spanish grant. Higuera was an heir of a soldier on the Anza expedition
W	Chabolla was an alcalde (mayor), not a Govenor; re-granted to Alviso
X	Milpitas means, "Little Corn Fields."
Y	Translated means, "Pig Spring." Purchased by Martin Murphy Sr. in 1846
Z	Translated means, "Shovel." White Rd. in San Jose goes through this land
AA	aka Rancho del Refugio; Mariano Castro was awarded half of the rancho in 1881
BB	Inigo was a mission Indian. There is a book on this individual
CC	See chapter 8
DD	Alviso was an heir of a soldier with the Anza Expedition; See Chapter 8
EE	
FF	Berryesa was a soldier at Presidio San Francisco; See notes for Chapter 8
GG	Robles purchased a large section from Pena in 1841
HH	Translated, "Corner of the Cats"
II	Widow of Rafael Soto, who died in 1839. Lawyers who helped secure patent got half of the land
JJ	Mostly in Santa Cruz County
KK	Another Rancho San Antonio was in the Berkeley area; Mesa was a Mexican soldier.
LL	Mesa led the expedition to put down Yoscolo. See note in Chapter 3
MM	
NN	Buelna helped oust Gov. Guitierrez. Alvarado thanked him with the grant. Title widely contested
OO	Not to be confused with the mission
PP	Son of Francisco Pacheco
QQ	Berryesa was a sergaent at Presidio San Francisco. Part of Almaden mines. Hotly contested title
RR	Spanish grant
SS	Ditto, son of Ignacio Ortega
TT	Heir of a member of Anza party
UU	In 1843 Alfred Chappell purchased El Rancho Solis.
WW	Grantees were mission Indians. J. Hoppe was the first US postmaster in San Jose
XX	Chabolla (Chaboya) was Pueblo San Jose police justice in 1834. Site of "Settlers War," 1861

Chart information is mostly from Arbuckle's *Santa Clara County Ranchos* and Bancroft. Some notes are from www.cagenweb.com/santaclara/landgrants.html. and from various other sources listed in the bibliography.

Appendix III: Plot Plan of the Pitman Farm

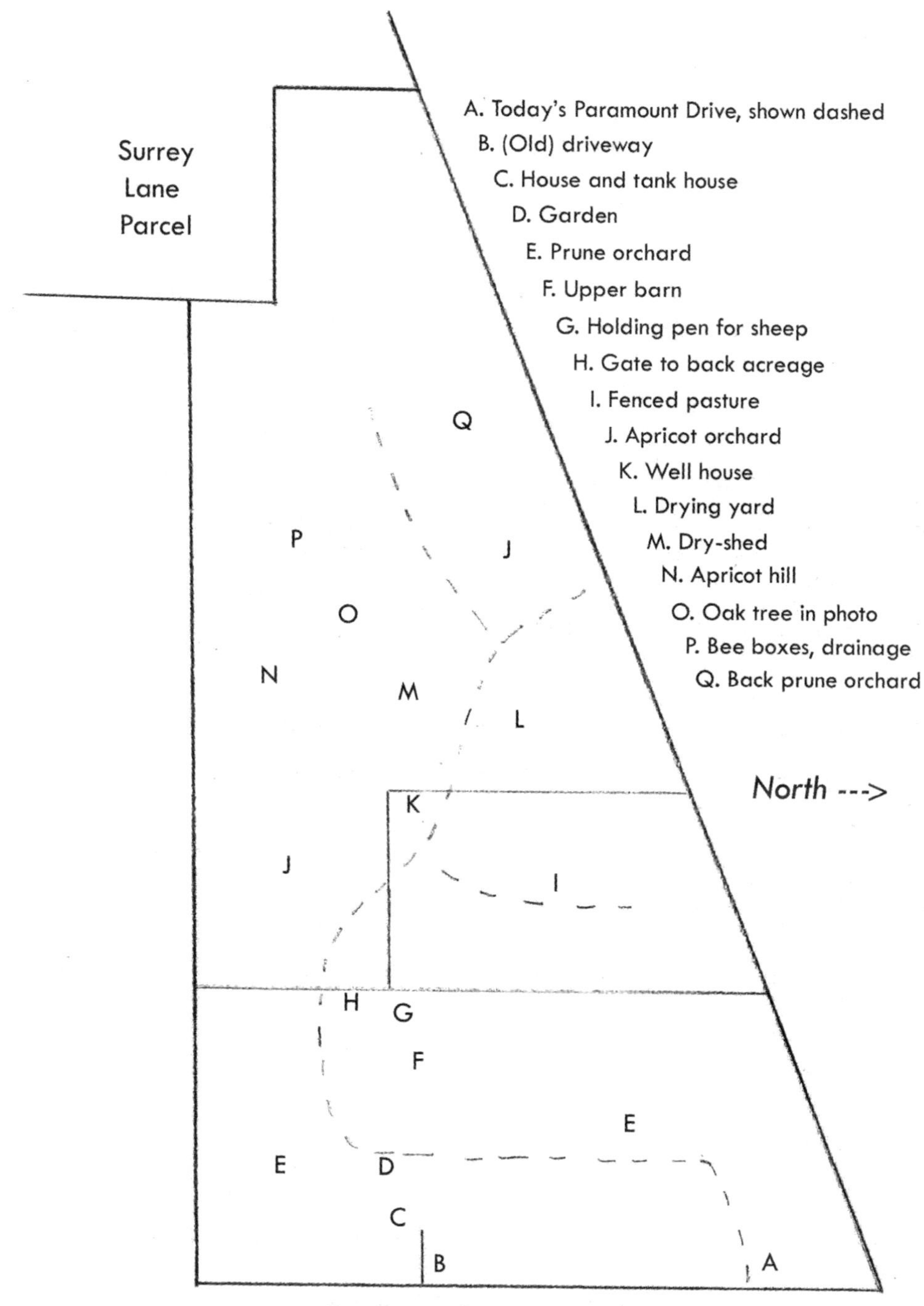

Photo Credits

Front Cover	Deborah L. Stanley
Front inside cover: Blanket of prunes	Deborah L. Stanley
Back inside cover: Apricots on tray	Tim Stanley
Back Cover	Tim Stanley

Map of Santa Clara Valley Vicinity	Deborah L. Stanley
Preface (orchard scene)	Courtesy of Calif. Room, San Jose Public Library
1. Beginnings:	
a. Acorns	Tim Stanley
b. Indian mortars	Deborah L. Stanley
2. Oak tree silhouette	Mari
3. Longhorn cattle	Jack Edmondson
4. Almaden mines	Courtesy, History San José
5. Tidal Wave:	
a. Quito diseño	Courtesy of the Bancroft Library, UC Berkeley
b. Los Gatos diseño	Courtesy of the Bancroft Library, UC Berkeley
6. Wheat field	Michael Creech
7. Enright's steam engine	Courtesy, History San José
8. 1876 Farm map	David Rumsey Map Collection
9. The Garden of the World:	
a. Tank house	Anna Rouse
b. Valley in bloom scene	Courtesy, History San José
10. Orcharding:	
a. Pear	Deborah L. Stanley
b. Orchard in bloom	Alberto Yáñez
c. Picking apricots	Courtesy, History San José
d. Prunes drying	Courtesy, History San José
e. Disking an orchard	Courtesy, History San José
f. Apricots to sulfur house	Courtesy, History San José
11. Picking strawberries	Simone Samuels
12. Market Challenges, Sunsweet:	
a. Working in a dehydrator	Courtesy of Sunsweet
b. Packing prunes	Courtesy of Sunsweet
13. Diary of a Cannery Worker:	
a. Cannery label	Courtesy, Martin Luther King Library, San Jose
b. Cutting pears	Courtesy, History San José
14. Old prune tree	Deborah L. Stanley
15. McGuire/ Pitman house	Courtesy of Dave and Tom Pitman
16. Pitman farm	Courtesy of Dave and Tom Pitman
17. Rooster	Kevin Morrison
18. Kleptomaniac fence scene	Deborah L. Stanley

19. Cow portrait	Danny Zveglic (allegrophotography.co.uk)
20. Apricots	Kathy Kimpel
21. Robert Pitman	Courtesy of Dave and Tom Pitman
22. Picking Prunes	Deborah L. Stanley
23.	(no photo)
24. Cleanup:	
a. Props in orchard	Deborah L. Stanley
b. Stacked fruit boxes	David Castanon
25. The "white" truck	Tim Stanley
26. Vegetable garden	Brad Johnson
27. Bucksaw	Tim Stanley
28.	(no photo)
29. Sheep	Courtesy of Richard van Yperen
30. Sheep dog	Dave Halley
31. Block and tackle	Andrew Fisher
32. 1943 prune picking crew	Courtesy of Ernie Kraule
33. Full boxes of prunes	Deborah L. Stanley
34. Model AA truck	David Berry
35.	(no photo)
36.	(no photo)
37. Prunes on tree	Deborah L. Stanley
38. Quito Creek bridge	Tim Stanley
39. Solitary farmer with shovel	stevensonphotography.com
40. Mustard bloom in orchard	Paul Miller
41.	(no photo)
Epilogue (fruitful branch)	Bill Bumgarner, http://friday.com/bbum

Appendix I: Modern Harvesting, (3) photos	Deborah L. Stanley
Appendix II: SCV Land Grants Chart, notes	Tim Stanley
Appendix III: Plot Plan of Pitman Farm	Tim Stanley

Bibliography

The Native Races of the Pacific States, Volume I, by Hubert Howe Bancroft, 1875

The Works of Hubert Howe Bancroft, Volumes XVIII - XXIV, 1890 (also titled: History of California, Volumes I-VII)

Situating Mission Santa Clara de Asis: 1776-1851, Documentary and Material Evidence of Life on the Alta California Frontier: A Timeline by Russell K. Skowronek with Elizabeth Thompson, Academy of American Franciscan History, Berkeley, CA , 2006

Pen Pictures from the Garden of the World, Edited by H. S. Foote, The Lewis Publishing Co., 1888

Reflections of the Past, An Anthology of San Jose, Heritage Media Corp., 1996

History of Santa Clara County, California, J. P. Munro-Fraser, editor; Alley, Bowen & Company, 1881

Clyde Arbuckle's History of San Jose, Smith and McKay Printing, 1985

Saratoga Story, by Vince Garrod, 1962

Saratoga's First Hundred Years, by Florence R. Cunningham, Valley Publishers, 1967

Santa Clara County Ranchos, by Clyde Arbuckle, 1968

Map of Santa Clara County Ranchos, by Ralph Rambo, 1968

History of Santa Clara County, by Eugene T. Sawyer, Historical Record Company, 1922

History of Los Gatos, by George G. Bruntz, Valley Publishers, 1971

Thompson and West Santa Clara County Farm Atlas, 1876, from the David Rumsey Map Collection

Passing Farms: Enduring Values, California's Santa Clara Valley, by Yvonne Jacobson, Wm Kaufmann, Inc., 1984

A Guidebook to California Agriculture, by the faculty and staff of the University of California, edited by Ann Foley Scheuring, University of California Press, 1983

An Overview of Ohlone Culture, by Robert Cartier with Laurie Crane, Cynthia James, Jon Reddington, and Allika Ruby, 1991

History of San Jose, by Frederic Hall, Bancroft and Co., 1871

Japanese Legacy, by Timothy J. Lukes and Gary Y. Okihiro, 1985

A History of the Chinese in California, Thomas W. Chinn, Editor, Chinese Historical Society of America, 1969

The Chinese Six Companies, by William Hoy, The Chinese Consolidated Benevolent Association, 1942

Chinatown San Jose, USA, by Connie Young Yu, History San José, 2001

Two Years Before the Mast, Richard Henry Dana, 1841

Decline of the Californios: A Social History of the Spanish-Speaking Californians, 1848-1890, Leonard Pitt, Ramon Gutierrez, University of California Press, 1999

Land in California by William Wilcox Robinson, University of California Press, 1948

Chicanos in California, A History of Mexican Americans in California, by Albert Camarillo, 1990

The Americans, Reconstruction Through the 20th Century, McDougal Littell, 1999

Historic Spots in California, by Mildred Brooke Hoover, Hero Eugene Renscb, and Ethel Grace Renscb, Stanford University Press, 1932

Archeology of El Presidio de San Francisco: Culture Contact, Gender, and Ethnicity in a Spanish-colonial Military Community, Barbara Voss, 2002 Ph.D. Dissertation, UC Berkeley

The Spanish and Mexican Adobe and Other Buildings in the Nine San Francisco Counties, 1776 to about 1850, Part VII, Hendry and Bowman, 1940, an-unpublished manuscript in the Bancroft Library, Berkeley.

The Sugar Beet in America, by Franklin S. Harris, 1919

The Great Wine Blight, by George Ordish, 1987

Web sites consulted:

Wikipedia
Online Archive of California
History San José
Silicon Valley History Online

California State Archives
US Surveyor General for California
National Archives
Unnamed book on the web site of the Contra Costa Water District

Santa Clara University
University of California
San Jose State University
San Jose Public Library

Santa Clara Research, The Valley of Heart's Delight
Moffett Field Museum
FMC Technologies
American Society of Mechanical Engineers

California State Parks
US Geological Survey
Santa Clara County Parks and Recreation
California Department of Parks and Recreation

Web Silicon Valley
Mount Diablo Interpretive Association
US Gen Web Project, CA Gen Web
San Jose Mercury News

Saratoga Historical Foundation
American Heritage.com
History Los Gatos
California Board of Food and Agriculture

Notes and References:

Major references are listed for some of the chapters. It should be noted however, that many chapters could not be completed without consulting many additional sources. This is especially true of chapters 1-5. See Bibliography.

Ch 1 Major refs: *Native Races of the Pacific States*, vol. 1; *Reflections*; Arbuckle

[1] *Reflections*, pg 14; Skowronek, pg 9; *An Overview of Ohlone Culture*. Names given to the different tribal groups vary widely in old writings.

[2] *An Overview of Ohlone Culture*

[3] *Native Races of the Pacific States*, vol. 1, chapter 4; Mission records, Santa Clara University; *An Overview of Ohlone Culture*

[4] The Spanish did not have the concept of separation of church and state. To be Spanish was to be Catholic. Two more missions were added later.

[5] "Indian(s)." For the sake of clarity, I have used "Indian" to describe the native peoples in a general way. It is the only designation that would not require the use of multiple terms in the text to describe the same peoples. The latter would cause confusion in the text.

[6] *The Works of Hubert Howe Bancroft*, vol. XVIII, ch 17

Ch 2 Major refs: Skowronek, *Reflections*, Bancroft, Arbuckle

[7] Skowronek, pg 166

[8] For a short time afterwards, there was a kind of labor-lending where the missions allowed certain Indians to work for the pueblo citizens, but this proved unworkable. The citizens of the pueblo then, having lost their labor source, were allowed to "recruit" Miwoks and Yokuts from the interior to do some of their labor (Skowronek, pgs 181, 192, 193). The Spaniards were familiar with the Yokuts and Miwoks because these peoples had found cattle much easier to hunt than elk and deer. Also, raids on the Spanish settlements by the Yokuts and Miwoks were not uncommon and on occasion were violent (Skowronek, pg 135). So there were some adjustments made on all sides.

[9] Skowronek, pg 82; Jacobson pg 29. The peacefulness of the Indians at Mission Santa Clara may also indicate a self-sacrificing life of most of the priests.

[10] "Serfs." The term "slaves" is too harsh—the Indians were not bought and sold. The term "converts," however, can be misleading. With perhaps good intention, the Indians were put under a kind of serfdom from which they were to "graduate" into full citizenship. This did not happen under Spanish rule and rarely, if ever, happened under Mexican rule. The idealism, however obviously impractical and previously proved unworkable, was convenient for political purposes at the time.

[11] The overwhelming majority of the "Spanish" in California were mestizos. By that time, Spain had been in the Americas for over two hundred years, and as few women emigrated from Spain, the men married Indian women.

[12] Wheat was by far the dominant crop. It was produced in abundance and at times was exported along with hides and tallow. Other crops listed consistently in the mission records were: barley, corn, beans, garbanzos, peas and lentils.

[13] Accounts of the extent of the enterprise vary greatly. *History of Santa Clara County*, 1881, pg 60, states that in 1825 Mission Santa Clara owned 74,400 head of cattle and 82,500 sheep. Jacobson, pg 27, states: "By 1830 there were 65,000 head of cattle within its control." However, mission records in the Santa Clara University archives for the same year (from Skowronek) state 7,000 cattle and 14,000 sheep. Mission records do not show the number of Mission Santa Clara cattle ever going over 14,000 in any year. Sorry, I could not solve this obviously important discrepancy. There are ample reasons why some might want to deflate the numbers (e.g., taxes), or inflate them (e.g., to lure new immigrants.) The rapid increase of the mission's cattle to over 4,000 head by 1792, and the fact that there were few ranchos until the missions were broken up, makes the mission's numbers for later years look too low. The limitation of water in the area and the historical record of the number of trading ships registered annually brings the higher figures into question. (See Bancroft, vol. XXI, ch 3; R.H. Dana's book.

[14] Skowronek, pg 187
[15] Skowronek, pg 281

Ch 3 Major refs: Bancroft, *Reflections*, Arbuckle
[16] Bancroft, vol. XX, ch 9. Most of the Hijar section is from Bancroft
[17] Bancroft, vol. XX, pgs 184, 249, 259
[18] Bancroft, vol. XX pg 260-279.
[19] Bancroft, vol. XX, pg 262
[20] Ejidos, or common lands, were a hold-over from the indigenous civilizations in Mexico. The Spanish and Mexican governments found it advantageous to keep the system in under-developed regions. With its many changes of government, Mexico has a long and colorful history of land reform and of the ejido system. In 1991 the system was officially eliminated.

[21] Embarcadero means "landing stage" or "pier."
[22] Skowronek, pgs 295, 302, 304; see Arbuckle also
[23] After the secularization of Mission Santa Clara and Mission San Jose, there were some short-lived Indian villages established around the bay. One of them was established by a man named Yozcolo (or Yoscolo) who helped himself to a large number of what had been mission cattle and horses, and established a community in the Los Gatos foothills. When it became evident that they could not support themselves, these people began raiding the nearby ranchos. In one of those raids, near New Almaden, an Indian guarding a cornfield and a ranchero were killed. The Mexican Army, aided by Indians from the mission area, then attacked Yozcolo's settlement in July of 1839. The battle is reported to have been fierce. A number in the attacking party and all of Yozcolo's followers were killed. Yozcolo's head was placed on a lance, paraded through Pueblo San Jose, and left on a pole near the front of the mission church for months as a warning to would be thieves and rebels. The leader of the attack, Juan Prado Mesa, was rewarded with a land grant of over 4,000 acres in the Los Altos area which became known as Rancho San Antonio. (Bancroft, vol. XX, pg 362; Skowronek pg 295-302) Accounts of Yozcolo vary greatly, but on this much the accounts agree.

[24] Three women later received US patents for Santa Clara Valley ranchos. (See Appendix II) One of particular note I will mention here, that is Juana Briones de Miranda. Unlike most Californio women who lived in this era, we have a significant written record of Juana Briones. She was born in 1802 in Villa Branciforte, now Santa Cruz, the daughter of a Spanish soldier and his wife. Her parents, as many "Spanish" in California at that time, were of mixed European, African, and Native American blood. Juana was married at age 18 to a Spanish soldier named Apolinario Miranda, who was stationed at Presidio San Francisco where the Briones family had recently relocated. The couple had eight children. Miranda was an abusive drunk, so to support herself and the children, Juana and her two sisters raised vegetables, milk cows, and cattle and sold the vegetables, milk, and hides to sailors who came ashore at Yerba Buena (now San Francisco). Juana was also a folk healer of the Spanish tradition and was known for her kindness toward sick and deserting sailors. Although illiterate (she signed her name with a cross), Juana Briones was a very capable businesswoman. With their earnings, she and her sisters purchased several pieces of real estate in the San Francisco Bay Area. In 1844, Juana petitioned the church for a legal separation from her husband, and from that time lived as a widow and dropped the name Miranda. In the same year, she purchased the Rancho Purisima Concepción from the two Indians it had been granted to and allowed them to remain on the land. The rancho was in today's Palo Alto and Los Altos. Señor Miranda died in 1849. Juana Briones died in 1889 at Mayfield, a small hamlet on or near the former rancho. Much legend has been propagated about this remarkable woman, but this much appears to be well documented. (Voss; Bancroft, vol. XIX, pg 730)

[25] See Appendix II; Bancroft, vol. XIX, pg 716

[26] Though the rancheros referred to themselves as "Spanish," nearly all were mestizos. *Chicanos in CA*, pg 7

[27] Tallow is animal fat. In those days, it was the primary ingredient used in both soap and candle making.

[28] *Reflections*, page 47

[29] *History of Santa Clara Co.*, 1881, pg 273; Historical plaque in Gilroy, CA

[30] Arbuckle, pg 24

[31] *Reflections*, pg 36

[32] Wikipedia, article on Tanning

[33] Bancroft, vol. XXI, pg 347

Ch 4 Major refs: Bancroft, *Reflections, Pen Pictures, History of SCC*, 1881

[34] Mexican government registers show the total population of California in 1845 as follows: Mexican citizens (Men of Reason), 6,900; foreigners, 680; former mission Indians, 3,180. The population was about evenly split in the northern and southern districts. The 6,900 figure includes foreigners who had become Mexican citizens and their children. (Bancroft, vol. XXI, pg 649) Other sources state the percentage of foreigners much higher. The number of foreigners by the end of 1846 overland migration season was very much higher.

[35] "Americans." In the larger sense, all those in the Americas are "Americans." When the Mexican government referred to "Americans," however, everyone knew

who they were referring to. In like manner, those from the US usually referred to the Mexican Californios as "Spaniards," and were equally understood.

[36] *Pen Pictures*, pg 38
[37] See Appendix II
[38] Bancroft, vol. XX, ch 6 and vol. XXI; *Reflections*
[39] *Land in California*, pg 68. The book has a map showing Central Valley ranchos.
[40] Bancroft, vol. XX, pgs 457, 458; vol. XXI, ch 1 gives a full account of Graham.
[41] CA State Parks web site article on Rancho Zayante; Bancroft. Historians greatly disagree on Graham.

[42] Bancroft , vol. XXI, pg 272
[43] Bancroft, vol. XXI, pg 274
[44] The Micheltorena War was not the only inter-Mexican armed conflict that took place partly on California soil. See Bancroft.

[45] *Pen Pictures*, pg 63
[46] Convicts were nothing new in California. California was already to some degree a penal colony. (Bancroft, vol. XX, pg 47.)
Further note: Micheltorena also enlisted a number of Americans to help him in the war.

[47] Hall, *History of San Jose*, pg 134, and *History of Santa Clara County*, 1881, pg 106, both state that they did declare California independent. Bancroft (vol. XXII, pg 507) however, treats this as a rumor started by Castro. Bancroft (pg 457) also states: "…the Californios, from beginning to end had no thought of throwing off their national allegiance to Mexico." However, a large body of evidence, including many other statements by Bancroft himself, contradicts this statement, at least as far as the northern aristocracy was concerned. What we do know is that their allegiance vacillated greatly and that they were continually posturing so that new regimes would not be too controlling over them.

[48] See Bancroft, vol. XXII, chapters 3-5
[49] *History of Santa Clara County*, 1881, pg 106
[50] *Reflections*, pg 60. Fremont would later become the hero of the American conquest. Some historians believe Fremont was set on conquest from the beginning, or that he was in fact sent for that purpose.

[51] *Pen Pictures*, pgs 64, 66. The California Republic lasted 26 days; the Republic of Texas, ten years.
[52] Most of the current Southwestern US was also part of the spoils of that war.
[53] Bancroft, in *Native Peoples of the Pacific States*, vol. 1, pg 370 quotes a source who states, "The New Almaden cinnabar mine has been from time immemorial a source of contention between adjacent tribes. Thither, from a hundred miles away, resorted vermilion-loving savages, and often such visits were not free from bloodshed." Mission Santa Clara records also refer to these conflicts.

Ch 5 Major refs: Bancroft, *Pen Pictures, History of SCC*, 1881, *Land in CA*

[54] Actually, many who came to California to farm in the early days of US occupation were recent arrivals to the US from various European countries.
[55] *Pen Pictures*, pg 35. Bancroft states the same, e.g., vol. XXIII, pg 532
[56] *Pen Pictures*, pg 32

[57] Another difference in culture is the way a Spanish person is named. A person of Spanish descent is referred to by both his father's last name, and then by his mother's maiden name. So with José Antonio Guerra y Noriega, Noriega is his mother's maiden name. In more modern times the "y" ("and") is dropped. Also, sometimes the mother's maiden name is dropped. So there can be confusion as to a person's last name when people from these two cultures interact. There was during the settling of the California land claims, and there is now.

[58] "Landholders" included all who held land—those of various nationalities who held lots in the pueblos, Mexican rancheros, American squatters or settlers, etc. Everyone had to pay the attorneys. Those who could not pay the attorneys commonly sold their interests to a speculator (often an attorney himself) who would then file a claim.

[59] Bancroft, vol. XXIII, ch 20. See also the SCV Land Grant Chart, Appendix II
[60] Purchasing land from a Californio before the claims were settled was no assurance that one would eventually receive a clear title. In a very real sense then, almost all California land purchases were by "speculators."

[61] Bancroft, vol. XXIII, ch 20. The provisions of the Treaty of Guadalupe Hidalgo stated that the Californio landowners would retain their land, but it seems that the newcomers were good at passing laws that nibbled at those provisions rendering them meaningless.
[62] *Saratoga's First Hundred Years*, pg 55; See also Appendix II, notes
[63] There were no formal grants for mission lands either (Ban., vol. XXIII, pg 562)
[64] *History of Santa Clara County*, 1922, Ch 3, pg 6
[65] Bancroft, vol. XXIII, pg 572
[66] Bancroft, vol. XXIII, pg 571
[67] Skowronek, pg 324
[68] *Land in California*, ch 13
[69] Bancroft, vol. XXIII, pg 577
[70] Bancroft, vol. XXIII, pg 544
[71] Bancroft, vol. XXIII, pgs 572- 578
[72] Skowronek, pg 20
Quote at the end of the chapter is from the Bible, Job, ch 7, v 17,18

Ch 6 Major refs: *Reflections, History of SCC,* 1881; Arbuckle, *Pen Pictures*
[73] Wheat was not new to the Valley. Mission Santa Clara had grown it profusely and at times even exported significant quantities of it.
[74] The wheat industry in California after the Americans came was very different than it had been during mission times. The combine harvester had come into use by the late 1840s, and California wheat farmers used harvesters pulled by teams of draft horses that cut a swath at least ten feet wide. Harvesting of wheat by hand

with a sickle, as was done in most of the world at that time, was virtually not done in California after the coming of the Americans.
[75] *History of Santa Clara County*, 1881, pg 23
[76] Arbuckle, *History of San Jose*, pg 140

Ch 7 Major refs: *History of SCC, Pen Pictures, Hist. of Los Gatos, Reflections*
[78] *History of Santa Clara County*, 1881 pg 37; *History of Santa Clara County*, 1922, ch 5, pg 3; USGS www.water.usgs.gov/mercury gives similar statistics.

[79] USGS www.water.usgs.gov/mercury. Interesting note: We look at the old mining practices and think that those people were irresponsible. However, perhaps we should look at our current practice. The list of currently used mercury-containing products is astounding. A very short list is: the new "green" florescent light bulbs (which contain mercury vapor), dental fillings, fungicides sprayed on crops and cut flowers, and preservatives used in some of our immunization vaccines. Oftentimes mercury is not listed in the ingredients of products because it is part of a compound. Such is the case of Thermisol, a common preservative used in vaccines. The list of ailments and diseases that are said to be a result of mercury poisoning is controversial, but all agree that mercury is an extremely dangerous toxic substance.

[80] *Reflections*, pg 90; *Pen Pictures*, pg 213
[81] *Pen Pictures* states 1860; *History of Los Gatos* states 1850s
[82] Garrod, pg 53
[83] *History of Los Gatos*, pg 20; Pen Pictures, pg 164
[84] Before the early 1860s, whale oil provided the fuel for lighting in a large part of the western world. By that time however, whale populations had been greatly diminished, and whale oil was very expensive. When distilling of petroleum oil into kerosene was perfected in the late 1850s, the modern petroleum industry was born. The US oil industry started in Pennsylvania in the late 1850s, and due to the shortage of whale oil, huge fortunes were made there. It is little wonder then, that the 1861 oil finds in the Los Gatos hills caused great excitement. Petroleum was a high tech industry of that time.

[85] History San José, photograph on file. The refinery burned down in 1897.
[86] *History of Santa Clara County*, 1881 pg 517. Engine patented in 1864.
[87] Arbuckle, *History of San Jose*, pg 140 states 200,000 acres

Ch 8 Major refs: Rambo, Arbuckle, *Pen Pictures*, Thompson and West, Bancroft, *Saratoga's First Hundred Years*
[88] See map at the end of the chapter
[89] See beginning of Chapter 3
[90] Bancroft, vol. XX, pg 737. Wife's name unknown.
[91] Bancroft, vol. XXI, pg 754, 755. Wife, Manuela Fernandez, one known child.
[92] See Appendix II ; aka Juan Ignacio Alviso. Wife, Margarita Bernal; 8 children.
[93] SF Genealogy web site; Bancroft, vol. XIX, pg 695
[94] Skowronek, pg 262; Bancroft
[95] José Manuel Alviso. Cunningham, pg 24,34. Wife, Manuela; six children

[96] Hendry and Bowman

Many Mexican rancheros were the recipients of bad legal council from American attorneys. Some sued each other. During the legal fighting over the land, the Alviso family was enticed to bring suit against another Californio ranchero who had settled next to them, the Berryessas. There was probably bad blood between the Alvisos and the Berryessas because the Berryessas had originally been granted Rancho Milpitas, but that grant was overturned, and the land was subsequently given to Alviso. (See Appendix II, note on Rancho Milpitas.)
One of the worst cases of the rancheros taking bad legal council also involved the Berryessa family. Their story is as tragic as any story of the demise of the Santa Clara Valley rancheros. An American attorney named James Jakes convinced them to abandon their adobe dwelling and settle on another portion of their land. Jakes then surveyed the rancho, parceled it out into quarter sections, and filed a claim for the Berryessa's home and surrounding acreage himself. Don Nicolas Berryessa sued Jakes but lost—and had to pay the court costs. The Berrryessas were then served with eviction papers, and instead of complying, armed themselves. Over time, Don Nicolas and three of his sons were driven insane by harassment from some of the squatters who were living on his land. A close relative, José Reyes Berryessa was lynched, apparently unjustly, and thereby lost his claim to the Almaden quicksilver mine. (Jacobson, pages 56, 65)

[97] Bancroft, vol. XX, pg 727
[98] *Pen Pictures*, pg 32; see also Skowronek pg 300
[99] The most extensive are *Pen Pictures* and *History of Santa Clara County*, 1881
[100] *History of SCC,* 1881, pg 652
[101] *Pen Pictures*, pg 618; *History of SCC,* 1881, pg 693
[102] *Pen Pictures*, pg 618. Cox had a few acres in fruit trees by the early 1880s.
[103] *Pen Pictures*, pg 521, 134, 526. There is some confusion concerning the locations of I.J. and J.R. Lovell's properties. Ann was the daughter of William Campbell, who came to CA in 1846 and built a sawmill on Saratoga Creek near the present Saratoga Springs Resort. Interestingly, he contracted with Manuel Alviso, then the owner of Rancho Quito, for the right to put in the mill. It was later determined that the mill was located far outside the boundaries of the rancho. (Cunningham, pg 25) The city of Campbell is named after William Campbell's son, Benjamin, who squatted on, and after an eighteen year legal battle received title to, what is now downtown Campbell. (*Reflections,* pg 54)

[104] *Pen Pictures*, pg 437; *History of SCC*, 1881, pg 667
[105] See middle of Chapter 5.
[106] Bancroft, vol. XIX pg701; Cunningham, pgs 59, 175, 177; *Pen Pict.*, pg 180
[107] Compare the shape of the Rancho Quito shown here with the diseño at the end of Chapter 5. The difference was likely determined by the written description that accompanied the diseño. The written descriptions referred to natural features of the land such as rock outcroppings, creeks, trees, etc.

Ch 9 Major refs: *History of SCC*, 1922; *Pen Pictures, History of SCC*, 1881; *California Agriculture, Reflections,* Jacobson

[108] American style wind pumps, or windmills, were patented by Daniel Halladay in 1854. Hallady's design differed from previous wind machines in that it automatically changed direction to face the wind and automatically controlled speed—huge improvements that changed farming the world over.

[109] *History of Santa Clara County* 1881, pg 23. An artesian well is one with enough pressure that the water flows up to the surface of the ground.
[110] *California Agriculture*, pg 208
[111] *Pen Pictures*, pg 173
[112] Garrod, pg 152
[113] Arbuckle, pg 155
[114] *History of Santa Clara County*, 1922, ch 12, pg 1
[115] Until the 1990s, there were two ways of developing new varieties of fruit trees: cross-pollination and grafting. Frequently a combination of the two methods is used. Even the best at developing new varieties, Luther Burbank, admitted that commercial success of a new variety was a matter of trial and error. Grafting method is discussed in Chapter 10.

[116] Jacobson, pg 90
[117] Property taxation almost certainly played a major part in the break-up of the wheat farms. The new California Constitution of 1878 changed how property was taxed, and within a decade most of the wheat farms in the Santa Clara Valley were gone. This could not have been a coincidence. However, it is very likely that even before the new constitution, taxation changes had resulted in the division of many of the Valley's wheat farms. Jacobson, pg 66, quotes an 1873 newspaper article stating that a property taxation system was just introduced. Unfortunately, I could not readily find a trustworthy source for the early (US) history of property taxation in CA, but the early development of the fruit industry in the Valley exactly coincides with the dates given here. Increased property taxation due to increased population had broken up the last of the ranchos, and later broke up the fruit farms. It is reasonable therefore, to assume that it played a significant part in the demise of the wheat farms as well.

[118] Jacobson, pg 89. Garrod (pg 156) adds, "A grower situated on suitable soil, with plenty of cheap water, good shipping facilities, a good supply of labor, in regard to quantity, quality and price; low taxes, free of debt, and raising a variety of crops could always make it. If one crop failed, he had another." I would add one more requirement here: the farmer maintaining good health. Several additional factors that determined if a farmer could make a living on his land are discussed in Chapter 10.

[119] Sugar was used as a preservative in canning because, like salt, it dehydrates the cells of organic matter, thereby retarding the growth of microbes which cause spoilage. Refining sugar from cane and beets, though done in the US from the 1830s, was not perfected until the late 1870s. (See *The Sugar Beet in America,* by Franklin S. Harris, 1919) Likewise, tin-covered wrought iron cans used for canning were commercially produced as early as 1812, but several decades were required before the idea was perfected. About the mid 1840s, the invention of

sheet steel was a huge technological leap. Early steel cans were soldered with lead-based solder and resulted in cases of lead poisoning. By the mid-1860s small machine-made cans were produced and made safer, but cooking to preserve most foods required about six hours, making canning uneconomical. The eventual development of efficient pressure cookers shortened cooking time significantly. From *Food Science* by N.N. Potter, J.H. Hotchkiss, 5th edition. Springer, 1999; *Food Processing Technology, Principles and Practice*, by P.J. Fellows, 2nd edition, Woodhead Publishing, 1999

[120] *History of Santa Clara County*, 1922. In chapter 12, pg 1, Sawyer states there were 23,900 farms in the county at that time, a figure which I do not think is possible. *Japanese Legacy*, pg 15, quoting census figures, states that there were about 6,000 farms at that time, and Jacobson states about 7,000 farms. The latter figures are probably more accurate. Perhaps the difference is in the definition of "farms."
Tonnage figures from different sources also vary considerably. Garrod states that 85,000 tons of prunes were produced in the Valley in 1900, making the 1921 figure here of 65,000 tons look low. Choosing one year to base statistics upon is not a good idea, but all my sources did so. Regardless, the figures stated give an idea of the size of the industry.

Ch 10 Major refs: personal experience, Garrod, *California Agriculture*
[121] See back cover. The Spanish brought the mustard to California. It is said that it was originally sown by the Spanish to mark a pathway between the missions.
[122] Garrod, pg 145
[123] *History of Los Gatos*, pg 98
[124] The moths lay their eggs at either the stem or blossom end of the fruit and when the larvae hatch they bore into the fruit. The typical "worm" in an apple, peach, or pear is a coddling moth larva.
[125] *California Agriculture*, pg 141. Most of the discussion of pears, peaches and their diseases is from *California Agriculture.*
[126] Bernard, Claude. *An Introduction to the Study of Experimental Medicine,* 1865. First English translation by Henry Copley Greene, published by Macmillan & Co., Ltd., 1927; reprinted in 1949; Encyclopaedia Britannica
[127] The schools serve two purposes: scientific research and practical training. Unfortunately, the objectivity of scientific research is easily compromised by the funding source(s). From the beginning of pesticide use, those endorsing the products have advised farmers and the public that, "It doesn't hurt people." With more than a hundred years of pesticide history behind us, some people—including many farmers—now find that harder to swallow. It is evident that the principle of pest eradication and the use of inexpensive agricultural chemicals have made possible the huge factory-farms we have today. It is equally evident, by the increasing number, volume, and potency of the chemicals used, that crop diseases are unimpressed with the chemical approach. In contrast, what is today called "organic farming" is based upon earlier principles of nurturing the soil and plants to avoid the weakened condition that allows the pests to thrive in the first place. There are benefits and dangers with both approaches. Some are long lasting. Unfortunately, these principles—killing and nurturing—seldom mix well.

Perhaps we should ask: is an occasional worm in an apple or a few aphids on our vegetables so undesirable that we would rather have our food laced with toxic substances and grown in sterilized soil that has left them deficient in nutritive value? We could also ask: how silent does spring have to become, how fostering of disease does our environment have to become, before we turn this around?

[128] California Agriculture, pg 273
[129] So when you wash off your produce under the kitchen faucet, guess what you're not washing off? (Use soap or a diluted solution of hydrogen peroxide.)

[130] Bean Ave. in Los Gatos is named after John Bean. Bean had previously invented a continuous flow turbine pump for windmills which was very successful. The Bean Spray Pump Company eventually became part of Food Machinery Corporation, which later became FMC. (*History of Los Gatos*)

[131] In 1901 the first California pesticide laws were passed (*Cal Ag*, pg 275). DDT, which had been used by the Allies both in Europe and in the Pacific during World War II to control malaria and typhus, was made available for agriculture in 1945. We are still paying for that decision. Since WW II, most of the pesticides used have been complex synthetic chemical compounds.

[132] The thrips stayed around for a few years, and many thought that was the end of prunes in the Valley. No treatment was effective for them, but after a few years, they had run their course, and most of the trees had survived. (Garrod, pg 156)

[133] Arbuckle, pg 158
[134] Grafting was probably developed in China about 2000 B.C. and was common in ancient Greece. No doubt the method was learned by observing natural grafts. The author has seen a natural graft in the Santa Cruz Mountains where one tree had fallen into another. The first tree was entirely uprooted, both trees were splintered, and the fallen tree had grown out of the other's rootstock. The trees were of different species.

[135] The Lester brothers were perhaps the largest prune growers in the world (Garrod). Toward the end of the era, they purchased or leased every prune orchard they could. They were also successful in talking IBM executives into leaving the prune orchard around IBM's Cottle Road facility in San Jose and allowing them to manage it. It was inexpensive landscaping and thankfully preserved some of the Valley's heritage.

Ch 11 Major refs: *Pen Pictures, Saratoga's First Hundred Years*, Chinn
[136] See end of Chapter 8
[137] *Saratoga's First Hundred Years*, pgs 54, 59, 175, 177
[138] There were two World's Fairs in 1939; the other was in New York City.
[139] *Reflections*, pgs 94-95
[140] Wikipedia, article on Phylloxera which sites *The Great Wine Blight* by George Ordish, 1987
[141] *Pen Pictures*, pg 185

[142] Before the Opium War of 1838-1842, Canton was the only Chinese port open to foreign trade. The Chinese living in that area, therefore, were far more familiar with the outside world than other Chinese. The Opium War was followed by several years of famine and other natural and political disasters in the Kwangtung area. Then gold was discovered in California. The gold rush served as a way out for many able-bodied young men, and they took it.

[143] *A History of the Chinese in California*, T. Chinn. Chinn gives a vivid account of the Chinese coolie trade, that is, the slave trade, which took place for more than two decades following the Opium War. He makes a clear distinction between that trade—mostly to Spanish and British colonies—and the early Chinese immigration to the US. The former were slaves—bought and sold under the same conditions as African slaves. The latter came on their own free will, and many paid their own fare. Those who did not have the money for the voyage paid off their fare by a credit-ticket system once they went to work in California. In the first few years of the gold rush, some were under contract for a specified length of time as indentured servants. Contracts for a specified length of time were soon abandoned when they were found to be unenforceable due to the prevailing anti-slavery sentiment in California at the time.

[144] *Pen Pictures*, pg 183; Chinn
[145] How independent these brokers were is a matter of historical controversy. Some historians have stated that an organization called the Chinese Six Companies controlled all Chinese labor in California. This was probably not the case, though to an outsider it may well have appeared so at the time, and the press often stated it as fact. The Chinese Six Companies, which later became known as the Chinese Consolidated Benevolent Association in San Francisco, acted as an intermediary for the Chinese in the US. The Chinese found it advantageous to govern themselves as much a possible and settle their own disputes. Not only did they suffer oppression from white Americans, and even from American law, there was also infighting, and not a few groups of hoodlums among the Chinese themselves. The hoodlums were organized. They controlled, and fought over, the opium and prostitution trades, terrorized Chinese business owners and citizens, and otherwise behaved just as criminals always have, and do today. It is no wonder then, that the Chinese empowered the Association to act as a kind of Supreme Court for the Chinese here. The Association also hired American lawyers to represent their interests in their adopted country.
The six "companies" were actually district associations that originally related to political districts in the Kwangtung area. Each association paid dues to the Six Companies in proportion to their membership. Chinese coming to California also paid a "departure fee" to the Six Companies before leaving China for California. These revenue sources paid for the services that the association provided for the Chinese community. (ref.: *The Chinese Six Companies*, pg 23. This book was produced by the Six Companies.) All these things being so, it is easy to see why Bancroft came to the conclusion that: "The Chinese Six Companies were really contractors and importers, although they attempted to pass themselves off as benevolent organizations. They governed and controlled with an iron hand, all the Chinese in the country." (vol. XXIV, pg 344) Bancroft may have exaggerated,

but even Chinn admits that late in the period an unsavory faction ruled the association. As always, what any group is comes down to the moral and ethical character of individuals.
The names of the original six companies were: The Sam Yup Company, Yeong Wo Company, Kong Chow Company, Ning Yung Company, Hop Wo Company and Yan Wo Company.
I found the book, *A History of the Chinese in California*, by T. Chinn, to be the most reliable source on the subject. Unfortunately it is rare and out of print.

[146] *Chinatown San Jose, USA*, pg 21
[147] *Reflections*, page 95
[148] *Japanese Legacy*, pg 32

Ch 12 Major refs: Garrod, *California Agriculture*
[149] There are many ways to manipulate a market. One example: today, if a newscaster on Channel 5 says spinach is bad for you, or that a hog in New Hampshire got sick, many people quit buying spinach or pork. Such a slight, unfounded comment can, and frequently does, greatly damage a market. Unfortunately, such seemingly innocent activities are often purposeful and timed to do the most damage. This is market manipulators at their worst.

[150] See Genesis, chapters 29-31
[151] How effective The Grange was politically is seen differently by historians. Compare *California Agriculture*, page 245; *The Americans*, pg 256
[152] "The fox guarding the henhouse" is closely related to "inadequate business skills." Several of the early and even later farmers' groups, recognizing their lack of business skills, formed associations and hired accountants and lawyers to handle their business affairs without putting adequate accountability safeguards into their agreements. So the accountants and lawyers got paid regardless if they produced for the association or not. Jacobson (pg 189) tells the story of her father who, as a member of an association, received a check for less than $2.00 a ton for his apricots. Granted this took place in 1931 during the Great Depression, but the point remains that there was not the necessary mechanism to ensure mutual benefit. (See chapter 33)

[153] Garrod, pg 162. The Sunsweet story comes largely from Garrod
[154] Changes in government policies, both state and national, played a large part in the success of the cooperatives. Sapiro could never have been successful before the large national monopolies were broken up, or before state laws became more favorable to agriculture. From the early 1860s, if not before, California agriculture struggled with imported insect pests. After some devastating disasters, methods of control and inspections were written into law. Likewise, controls were needed on the human machinations that could, and did, just as surely devastate an industry. The Cooperative Marketing Act, passed by the state legislature in 1895, was perhaps the first significant step toward rectifying some of the abuses. The California Board of Food and Agriculture, a state agency, was eventually formed in 1919 to deal with both problems. It was the result of more than four decades of wrestling with agricultural issues. See Garrod, CBFA website.

[155] *History of Santa Clara County*, 1922, ch 12, pg 1

Ch 13
[156] All names in this chapter are entirely fictitious, as are all of the main character's thoughts.

Ch 14 Major refs: *Reflections; The Americans, Reconstruction to 20th Century*
[157] From "Government Issue"
[158] *Reflections*, pg 183. Other cities also joined in on the land grab; for example Los Gatos annexed some of the Cambrian area at that time.
[159] See middle of chapter 41

Ch 15 Major refs: *Pen Pictures, Saratoga's First Hundred Years*
[160] *Japanese Legacy* and some other sources state that the Chinese accounted for nearly 50% of the farm labor force in the Valley during this period. However, at the height of the Chinese labor period, in the 1870s and 1880s, the Chinese did not exceed 8% of the SC Valley's population (Chinn, pg 21, census table). Since the primary industry in the Valley was agriculture, the 50% figure is misleading. If one does not count the farm owner and his family, and only counts the lowest paid bulk labor, the 50% figure is probably about right. (See Chapter 32) Chinese laborers no doubt performed a high percentage of the most arduous, dangerous, and low-paying work, not only in the Santa Clara Valley, but also in the state as a whole. California could not have been built up so quickly without them.

[161] *Pen Pictures* bios; *Saratoga's First Hundred Years*. Rice was born in 1821.
[162] *Pen Pictures* bios; *Saratoga's First Hundred Years*. McGuire was born in 1850.
[163] *Saratoga's First Hundred Years*, chapter 23
[164] Garrod, page 164

Ch 16
[165] Chapters 16-19 are mostly from interviews with Dave Pitman. I have left Dave's flavor in recounting the stories.
[166] There was a buffalo on the back of the nickel from 1914-1938.

Ch 21 Quote at the end of the chapter is from the Bible, I Cor. ch 6, v12
Ch 22 [167] See chapter 10

Ch 23
[168] Although some despicable acts were committed by some individuals, the large immigration of people of Japanese ancestry into the Santa Clara Valley after WW II may be ample testimony that during that difficult time most of their kinsmen had been treated with respect by the other citizens of the Valley.

Quote at the end of the chapter is from the Bible, Matthew ch 7, v12

Ch 25 Quote at the end of the chapter is from the book of Proverbs, ch 27, v7
Ch 26 Quote at the end of the chapter is from the Bible, S of S ch 2, v15

Ch 27
[169] Another point well worth mentioning is that this individual obviously did not pay for his own medical bills.

Ch 29
[170] "As white as snow." Note worth pondering: White is the absence of color. Snow being pure white we can understand—it comes from the heavens. The inside coat of a creature on this dusty earth being pure white is a lot more difficult to understand. Yet this is the picture given in the Bible of the forgiveness of sins by God. Isaiah the prophet spoke of this in the first chapter of his book. He used the terms "white as snow" and "like wool" to describe God's cleansing of our sins if we but confess them to Him. See Isaiah ch 1; I John ch 1.
Quote at the end of the chapter is from Isaiah, ch 53

Ch 32 Major refs: Garrod, *California Agriculture*
[171] *California Agriculture*, page 16
[172] By the 1920s, California had an efficient system of farm worker migration. With the addition of the winter crops—olives, citrus, and winter vegetables—nearly year round work was available for many workers who traveled up and down the state following the crops. Much has been written about the abuses of migrant farm workers, but there are also a large number of stories of mutually beneficial relationships. Many workers had (and have) good, personal relationships with farmers in different areas of the state and were (are) quite happy with their arrangements. The common thread in all the stories of these relationships is the subject of Chapter 33.

Ch 33
[173] The safeguard to the worker in the piecework system was that if the farmer was not paying the going rate or otherwise taking care of his workers, the workers would go elsewhere. If a farmer's fruit was small, if he did not get the weeds out of his orchard or groom it properly, if he did not supply good equipment or otherwise provide a suitable workplace for the harvesters, they would pass him by. The workers were free agents, and word got around. There are plenty of stories of farmers who didn't get their crops picked, just as there are stories of workers who were not treated fairly.

Ch 34
[174] See Chapter 41, 2nd paragraph.
Ch 36 Quote at the end of the chapter is from Psalm 32, v1

Ch 38
[175] A hand pipe threader is a bar with thread cutting dies on the end. The dies are slipped over the end of the pipe and the threads are cut by pushing down on the lever or bar to rotate the cutting die around the pipe. Like using a pipe wrench.

Ch 41
[176] Garrod, page 146 and conversation with George Mackenzie, 2009

Epilogue Quote at the end is from the Bible, Phil. ch 2, v7

By the Same Author:
Letters to My Feathered Friends
Observations, Meditations and Thanksgivings

A collection of seventy-six bird poems and stories with complementing color photographs.
142 pages, soft cover

For a copy of *Letters to My Feathered Friends* by mail, send check or money order to: 2 Timothy Publishing, P.O. Box 53783, Irvine, CA 92619. Shipping and handling are included in the amount below.
Total cost: $21.00 (CA mailing addresses add $1.50 sales tax)

You may also order online at www.2timothypublishing.com